SPSS® STUDENT MANUAL AND WORKBOOK

JAMES J. BALL
Indiana State University

to accompany

THE TRIOLA STATISTICS SERIES:

Elementary Statistics, Tenth Edition

Elementary Statistics Using Excel, Third Edition

Essentials of Statistics, Third Edition

*Elementary Statistics Using
the Graphing Calculator*, Second Edition

Mario F. Triola
Dutchess Community College

PEARSON

Addison
Wesley

Boston San Francisco New York
London Toronto Sydney Tokyo Singapore Madrid
Mexico City Munich Paris Cape Town Hong Kong Montreal

Reproduced by Pearson Addison-Wesley from electronic files supplied by the author.

Copyright © 2007 Pearson Education, Inc.
Publishing as Pearson Addison-Wesley, 75 Arlington Street, Boston, MA 02116.

ISBN   0-321-36911-4

2 3 4 5 6 BB 09 08 07 06

Preface

This SPSS Manual was written to demonstrate how to solve statistical problems using *SPSS 14.0 for Windows*®. It is intended to be used in combination with *Elementary Statistics,* 10/e or *Essentials of Statistics,* 3/e both by Mario Triola (Pearson Addison-Wesley, 2007).

This manual begins with an introduction to SPSS that is intended to introduce the student to the basic commands and menus. Chapters 1 – 14 follow the development of topics in *Elementary Statistics,* 10/e making it very easy to use as a supplement. Each section begins with a paragraph summarizing the statistical topics to be discussed followed by a discussion of how to use the relevant SPSS procedures. Each procedure is demonstrated by doing examples from the textbook.

SPSS 14.0 for Windows is a research quality statistical software program that is easy to learn and use. Using a statistical program such as SPSS allows the student to explore problems and concepts in more detail and to gain a better understanding of the principles and concepts of statistics.

The Student Version of SPSS is a very affordable alternative to the commercial software. The Student Version is limited in the number of variables and cases it can process and lacks the ability to read command syntax. In most instances, students will never notice these limitations. For more about the SPSS program, contact SPSS Inc. at 233 S. Wacker Drive, 11th floor, Chicago, Illinois 60606 or visit their Web page online, http://www.spss.com.

First, I would like to congratulate Mario Triola for writing excellent statistics textbooks and for permitting the use of many examples from them. I would also like to thank the people at Addison-Wesley for their kind support in the creation of this manual. I would especially like to thank Christine O'Brien for her advice and patience and Joe Vetere for all of his expertise with graphics and illustrations.

James J. Ball
Department of Mathematics and Computer Science
Indiana State University

May 2006

Contents

Chapter 0

Introduction to SPSS

Chapter 0   Introduction to SPSS

The purpose of this manual is to teach you how to solve statistical problems using the statistical software program **SPSS 14.0 for Windows**. SPSS is a research quality statistical software program that is easy to learn and use. SPSS has versions available for IBM compatible and Macintosh personal computers as well as several mainframe systems. This manual is specifically designed to work with **SPSS 14.0 for Windows**. If you are using a different version of SPSS or using SPSS on a computer not using Windows, you will still find this manual helpful since the different versions of SPSS are very similar.

This manual is designed to accompany *Elementary Statistics*, 10/e or *Essentials of Statistics*, 3/e both by Mario F. Triola[1]. This manual is structured to help you work through the examples and homework that appear in *Elementary Statistics* and *Essentials of Statistics*. The best way to learn anything is to do it. You should have your textbook with you and be at a computer with SPSS running while you read this manual.

This chapter is a brief introduction to SPSS. The rest of the chapters in this manual correspond to the chapters in your textbook. This manual has been written with *Elementary Statistics* as the primary reference since all the material in *Elementary Statistics* is included in *Essentials of Statistics*. It is best if you have read the corresponding chapter in your textbook prior to beginning the chapter in this manual. The current chapter may be begun immediately because it does not require any statistical knowledge.

This chapter discusses some of the basic menus and commands in SPSS. You will learn how to create, edit, and save data files. Further, you will learn how to get help from SPSS and how to print. We assume that the reader is familiar with the basics of using Windows including the use of a mouse, and being familiar with menus and dialog boxes. In this manual, SPSS commands and actions will be shown in boldface font (e.g. choose **File**, then choose **Open...**). Key terms will be shown in boldface fonts (e.g. **Case** or **Data Editor**) and variable names will be shown in boldface and italics (e.g. *Var1* or *Salary*).

SPSS 14.0 is based on point-and-click technology. In this manual, it is understood that "choose **File**" means move the cursor over the word **File** and then click the left mouse button. Sometimes, after choosing something (e.g. choose **File**) a new list of choices will appear. In this manual, the shorthand notation "choose **File > Exit**" will be used to indicate, move the cursor over the word **File**, click the left mouse button, then move the cursor over the word **Exit**, and click the left mouse button again.

---

[1] Published by Addison-Wesley, Boston.

Section 0-1   Starting SPSS

There are many different computer systems (IBM compatible PC, Macintosh, mainframe computers, etc.) that SPSS runs on, it might be necessary to ask your computer administrator how to start the SPSS program. To start SPSS 14.0 for Windows on a machine running the Microsoft Windows 2000 operating system, click the **Start** button and then move the mouse to **Programs.** After a few seconds a list of the computer programs on your machine will appear. Browse the list of programs that appear until you locate SPSS for Windows and choose **SPSS 14.0 for Windows**. When **SPSS 14.0 for Windows** starts the window shown in Figure 0 - 1 will appear on your computer screen.

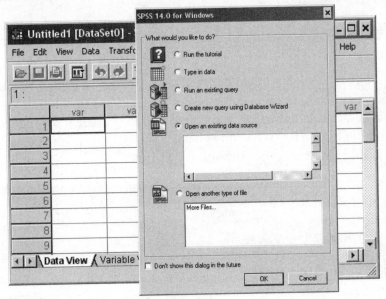

Figure 0 - 1

If you are unfamiliar with SPSS, it is best to become acquainted with SPSS by going through the tutorial. The tutorial will take about an hour; you can cancel the tutorial and restart it later should you change your mind. To begin the tutorial, choose the bullet for **Run the Tutorial. C**lick the **OK** button and then follow the instructions in the **Tutorial**.

If you want to enter data into a new SPSS data file, choose the bullet for **Type in data**. If you want to open a file on your computer, choose the bullet for **Open an existing data source**. The box below "Open an existing data source" lists some recently used SPSS data files. The box you see may different than shown in Figure 0 - 1. It may have a list of SPSS files depending on what files have been recently opened in SPSS on your computer. Clicking on a filename in the list and then clicking the **OK** button will quickly open the selected file. We will not discuss the other options (Run an existing query, Create a new query using Database Wizard, or Open another type of file) in this manual, as it is unlikely you will need to use them.

When you are ready, choose the bullet for **Type in Data**. Along the top of the SPSS window is the menu bar (Figure 0 - 2) where all the commands used in this manual can be found. The menu bar has eleven menu items (**File, Edit, View, Data, Transform, Analyze, Graphs, Utilities, Add-ons, Window, Help**) above a row of icons that are shortcuts for many of the common commands, such as opening or saving data files and printing. Each of these menu items has sub-menus that can be accessed by choosing the menu item.

Figure 0 - 2

Choosing a sub-menu item will usually open a **dialog box** but sometimes will open a sub-sub-menu. Dialog boxes are used to select variables and options for analysis. Each dialog box associated with a statistical procedure or chart (for example Figure 0 - 3) has three basic components: **a source variable list**, a **target variable list**, and some **command buttons**.

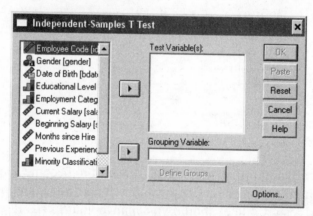

Figure 0 - 3

The **source variable list** is a list of variables in the data file that can be used by the procedure. The left-most box in Figure 0 - 3 contains the source variable list for this procedure. Only variable types that are allowed by the procedure are displayed in the source variable list. **Target variable list(s)** are one or more lists indicating the variables chosen for the analysis. Both the **Test Variable(s)** box and the **Grouping Variable** box in Figure 0 - 3 are target variable lists.

Measurement Level	Data Type			
	Numeric	String	Date	Time
Scale		n/a		
Ordinal				
Nominal				

Figure 0 - 4

SPSS uses different icons to provide information about the variable type and measurement level. Notice that by referring to Figure 0 - 4 that the variable *gender* is nominal string data, ***date of birth*** is scalar date data, ***educational*** **level** is ordinal numeric data, and ***current salary*** is scalar numeric data (see Section 1-1 of this manual for more information about level of measurement).

The **variable paste** ▶ button is used to paste a variable from the source variable list to the target variable list. The **command buttons** (**OK, Paste, Reset, Cancel, Help, Define Groups…,** and **Options…**) are buttons that instruct the program to perform an action, such as run a procedure, display Help, or open a sub-dialog box to make additional specifications. Buttons that are grayed out (e.g. **OK, Paste**, and **Define Groups…** in Figure 0 - 3) are disabled and cannot be used. These buttons will be available once variables have been selected into the target variable list.

There are six standard buttons (**OK, Continue, Paste, Reset, Cancel,** and **Help**) in most dialog boxes. After you select your variables and choose any additional items in the dialog box, click the **OK** button and the dialog box will close and run the procedure. Pressing the **Enter** key is equivalent to clicking the **OK** button. In sub-dialog boxes, there is a **Continue** button. The **Continue** button is similar to the **OK** button except it does not run the procedure. Clicking the **Continue** button will save any changes you have made, close the sub-dialog box, and return to the original dialog box. Clicking the **Reset** button will deselect any variables in the target variables list(s) and reset all specifications in the dialog box and any sub-dialog boxes to their default state. Clicking the **Cancel** button will close the dialog box and cancel any changes in the settings since the last time the dialog box was opened. Clicking the **Help** button will open a context-sensitive dialog box that contains information on the current dialog box. Clicking with the right mouse button (right-clicking) in many of the individual dialog box controls will open a context-sensitive help window containing information about that control.

Section 0-2   Opening an Existing Data File

Many data files already exist on your computer. The eighteen SPSS data files listed in **Appendix B: Data Sets** of your textbook are saved on the data disk, which comes with the textbook. Another fifty-six data files come with **SPSS 14.0 for Windows**. In this section, we will learn how to open these existing data files. SPSS can open data files stored in many different formats. For example, it can open data files created by Microsoft Excel, dBase, Lotus 1-2-3, Systat, and SAS. SPSS can also open data files that are stored in SYLK (symbolic link) format that is used by some spreadsheets. It can also read data from text files.

The **Employee data.sav** data file comes with SPSS. This data file has information on *474* employees hired by a midwestern bank between 1969 and 1971. The bank was subsequently involved in EEO litigation. For additional information about this data file, see "Statistical bases in the measurement of employment discrimination" by H. V. Roberts (1979). In: *Comparable Worth: Issues and Alternatives*, E. R. Livernash, ed., Washington, D.C.: Equal Employment Advisory Council.

We begin by opening this SPSS data file. To open an existing SPSS data file, choose **File > Open > Data…** (i.e. choose **File** from the menu, then choose the sub-menu item **Open,** next choose the sub-sub-menu item **Data…**) and the **Open File** dialog box (Figure 0 - 5) will open. The dialog box shows a list of folders and SPSS data files.

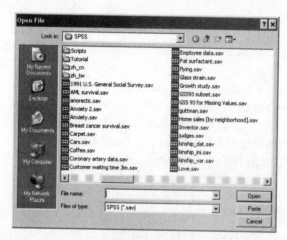

Figure 0 - 5

By default SPSS data files (.sav extension) are displayed in the **Open File** dialog box. Files of other data formats can be displayed using the **Files of type** box. Click the **down arrow** button at the right of the **Files of type** box to see the different formats that SPSS can open. In this case, **Employee data.sav** is stored in SPSS format therefore we do not want to change this option. If you select a different format then no data files will be showing in the window since this folder contains only SPSS files.

The data files are listed alphabetically. If **Employee data.sav** is not showing in the dialog box, scroll through the data files by clicking the ▶ button until **Employee data.sav** is showing. Open the data file named **Employee data.sav** by clicking on its name and then clicking the **Open** button.

The **Employee data.sav** data file will appear in the **Data Editor** (Figure 0 - 6). If the **Data Editor** opened in the **Variable View** then click on the **Data View** tab near the bottom of the window. SPSS is now ready to perform statistical calculations on this data.

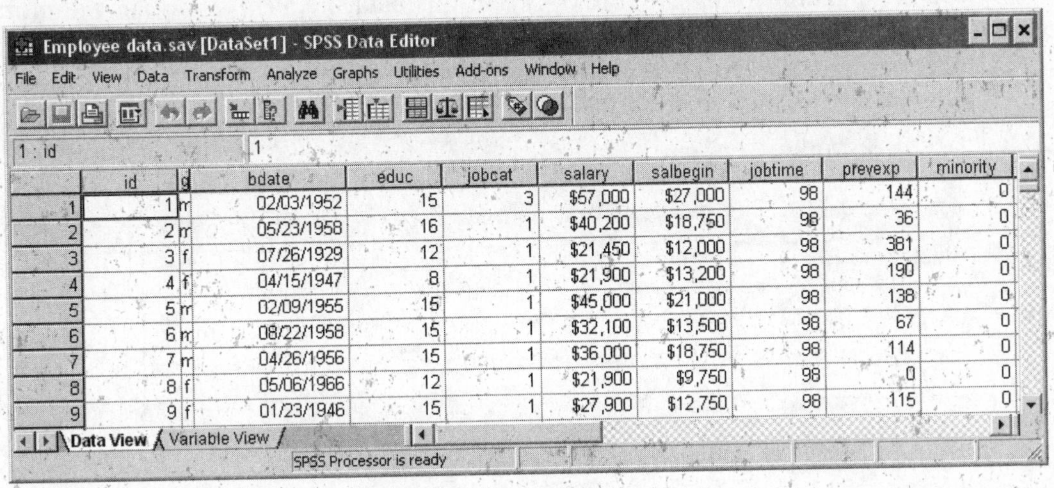
Employee data.sav [DataSet1] - SPSS Data Editor
File Edit View Data Transform Analyze Graphs Utilities Add-ons Window Help
1 : id 1
id g bdate educ jobcat salary salbegin jobtime prevexp minority
1 1 m 02/03/1952 15 3 $57,000 $27,000 98 144 0
2 2 m 05/23/1958 16 1 $40,200 $18,750 98 36 0
3 3 f 07/26/1929 12 1 $21,450 $12,000 98 381 0
4 4 f 04/15/1947 8 1 $21,900 $13,200 98 190 0
5 5 m 02/09/1955 15 1 $45,000 $21,000 98 138 0
6 6 m 08/22/1958 15 1 $32,100 $13,500 98 67 0
7 7 m 04/26/1956 15 1 $36,000 $18,750 98 114 0
8 8 f 05/06/1966 12 1 $21,900 $9,750 98 0 0
9 9 f 01/23/1946 15 1 $27,900 $12,750 98 115 0
Data View Variable View
SPSS Processor is ready

Figure 0 - 6

The data displayed in your Data Editor may look different from those shown here since there are several different options available to customize the view. The next section discusses some of these options. Before doing any calculations let's explore the features of the **Data Editor**.

Section 0-3 The Data Editor

The **Data Editor** provides a convenient, spreadsheet-like display for creating and editing data files. The Data Editor window opens automatically when you start a session. SPSS provides two views of the data file in the **Data Editor**, the **Data View** and the **Variable View**. The variables and data values (or defined value labels) are displayed in the **Data View**. The **Variable View** displays variable definition information, including defined variable and value labels, data type (for example, string, date, and numeric), measurement scale (nominal, ordinal, or scale), and user-defined missing values.

Cases, **variables**, and **cells** are basic concepts in SPSS data files. In this data file, each **case** (each row in the data file) corresponds to a different employee. A case (sometimes referred to as an observation) is an employee, a person, or anything that is being studied. **Variables** (each column) are different pieces of information collected on the cases.

The **Employee data.sav** data file has *10* variables and *474* cases. The variables are the ten different measurements *id, gender, bdate, educ, jobcat, salary, salbegin, jobtime, prevexp,* and *minority* made on each employee. Each variable-case combination denotes a **cell** in the data file. For example, the intersection of case 4 and variable 3 (*bdate*), denotes the cell which has the date 04/15/1947 in it. There are *4,740* cells (*474* cases · *10* variables) in this data file.

Click the **Variable View** tab near the bottom of the **Data Editor** (Figure 0 - 6) to see the **Variable View** (Figure 0 - 7) for this data file. This view shows information about each of the variables. If you click the **Data View** tab, the **Data Editor** window will return to the **Data View**.

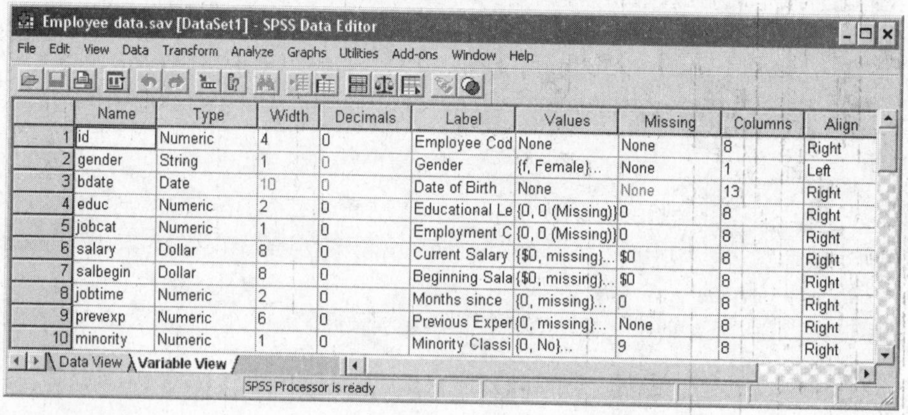

Figure 0 - 7

SPSS limits variable names to eight characters or less and the name cannot include any spaces or special characters. This can make it difficult to tell what the variable is describing. For example, consider what the variables *id* or *jobtime* might be measuring in **Employee data** data file. To solve this problem SPSS allows each variable to be associated with a label. **Labels** can be up to *256* characters long and can therefore be more descriptive than variable names. The column named **Label** in the **Variable View** shows the labels for the variables. SPSS uses the variable labels in the output from statistical procedures.

Variables in SPSS can have eight different **data types**. The variables *id*, *educ*, *jobcat*, *jobtime*, *prevexp*, and *minority* are all numbers with no decimal points showing and have data type, *Numeric*. The variable *gender* is character data and has data type, *String*. The variable *bdate* is a date and has data type, *Date*. The variables *salary* and *salbegin* are currency and have data type, *Dollar*.

Click in the cell that reads **Date**, in the row for *bdate* and the column labeled **Type**. Click on the dots icon ⬚ that appears and the **Variable Type** dialog box (Figure 0 - 8) will open.

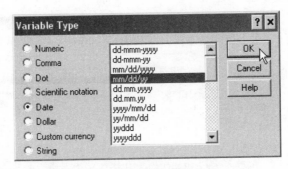

Figure 0 - 8

Click on **mm/dd/yy** as the variable type and then click the **OK** button. SPSS will now use the two-year date format. Notice that in **Data View**, the third case of *bdate* shows ********.   The asterisks are indicating that the date is out of range. SPSS automatically defines the range of years for date-format variables and date functions that use a two-year data instead of a four-year date. The automatic setting defines a 100-year range beginning 69 years before the current year and ending 30 years after the current year. This data value was originally entered as 7/26/1929, but SPSS treats 7/26/29 as July 26, 2029. This is because 1929 is 77 years prior to 2006 (the current year). We can undo the change we have just made by choosing **Edit > Undo Modify Variables** from the **Edit** Menu. The **mm/dd/yy** format requires 8 columns whereas the **mm/dd/yyyy** format requires 10 columns to show it properly. Scroll to the right in the **Variable View** window (Figure 0 - 7) by clicking the ▶ button until the column labeled **Columns** is displayed. The number in the column shows the width of the column in characters that will be displayed in the **Variable View** and can be changed to match the selected format requirements for *bdate*.

Notice that in Figure 0 - 6 the column displaying the variable *gender* is very narrow. Change the column width for gender to *8*. Now the variable *gender* will be displayed using eight columns.

Just as a variable can have a label that identifies the variable, **value labels** up to *60* characters long, can be assigned to the values of a variable to help identify the values. The variables *gender*, *jobcat*, and *minority* use **value labels**. To see how the **value labels** for *jobcat* are defined click in the cell that reads **{0,0 (Missing)}...** in Figure 0 - 7. This cell is in the row for *jobcat* and the column corresponding to **Values.** Click on the dots icon ⦙ that appears and the **Value Labels** dialog box (Figure 0 - 9) will open.

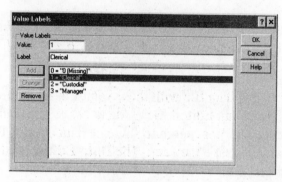

Figure 0 - 9

Click on **1 = "Clerical"** to observe how the value label is defined. This dialog box can be used to add, change, or remove value labels. We do not want to change the value labels, therefore click the **Cancel** button to return to the **Variable View**.

Change to the **Data View** by clicking on the **Data View** tab. To display the **Value Labels** in the **Data View**, choose **View > Value Labels** from the **Data View** menu. Compare the current **Data View** (Figure 0 - 10) with original **Data View** (Figure 0 - 6). This setting is a toggle, which means that choosing **View > Value Labels** again will display the data values.

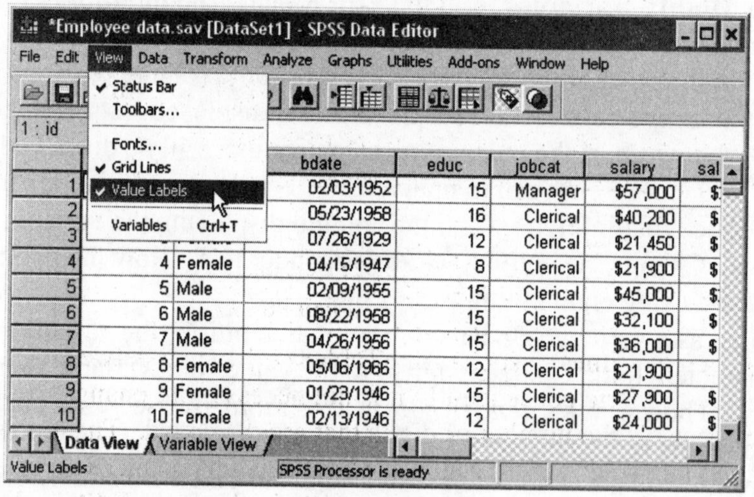

Figure 0 - 10

Section 0-4   Typing in a New Data File

It is often necessary to enter data into a data file in SPSS before doing any analysis. Consider the data in Problem 17 of Chapter 3.2 of the textbook, **Customer Waiting Times.** When investigating waiting times required for customer service, the following results (in minutes) are obtained. The times in minutes are given in Table 0 - 1.

Jefferson Valley Bank:	6.5	6.6	6.7	6.8	7.1	703	7.4	7.7	7.7	7.7
Bank of Providence:	4.2	5.4	5.8	6.2	6.7	7.7	7.7	8.5	9.3	10.0

Table 0 - 1

To create a new data file with these values in it, choose **File > New > Data.** If there is data in the Data Editor, SPSS might ask if you want to **Save the contents of data editor**, click the **Yes** button to save the data or click the **No** button to clear the **Data Editor** and not save the data. The **Data Editor** is empty and SPSS is ready to have new data entered.

This data has *20* cases (*10* times from Jefferson Valley Bank and *10* times from Bank of Providence) and two variables, ***minutes*** and ***bank***. Choose the **Variable View** tab if the **Data Editor** is in **Data View**. In this view, the first column is labeled **Name**. To create a new variable named ***minutes***, you need to type *seconds* into the first row of the column labeled **Name**. When you press the enter key, the new variable *seconds* is created. SPSS automatically sets the **Type** of new variables to *Numeric*. SPSS also automatically sets **Width** to *8* and **Decimals** to *2* for new variables. SPSS sets the level of measurement to **Scale** (see Section 1-1 of this manual for more information about level of measurement). SPSS does not automatically define a **Label** or **Value Labels** for new variables. Figure 0 - 11 shows how the **Data Editor** records this information.

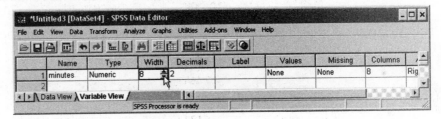

Figure 0 - 11

Notice the times in seconds in Table 0 – 1 are all between 4.2 and 10.0. The values are numeric consequently it is not necessary to change the **Type** setting. **Width** controls the number of digits that are displayed. The data values might be displayed incorrectly if the value for **Width** is too small. Regardless of the choice for **Width,** the data values are stored correctly in the **Data Editor**. **Width** for this data should be at least four; use **Width** 5. To set the **Width** to the value 5, click in the cell for **Width** and some arrows will appear (see Figure 0 - 11). Click the arrows up or down to change the value. Alternatively, you can simply type the number 5 into the cell. **Decimals** controls the number of decimals displayed in the **Data Editor**. Since the data values have one decimal place, use **Decimals** 1. This data format is referred to as *Numeric5.1*.

Click in the cell for **Label** and type in *Minutes required for service* to create an appropriate label for this variable. The label can be up to *256* characters long.

The next variable is an indicator of whether the bank is Jefferson Valley Bank or Bank of Providence. Create the variable ***bank*** by typing *bank* into second row of the column labeled **Name**. When we enter the data for *bank*, we will use *J* to indicate that the bank is Jefferson Valley Bank and *P* to indicate Bank of Providence. The values *J* and *P* are characters; therefore, the data **Type** should be *String*. To change the variable type of the variable *bank*, click in the cell for **Type** in the row for *bank*. Click the ▦ icon and the **Variable Type** (Figure 0 - 12) dialog box will open.

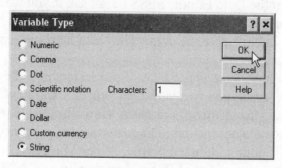

Figure 0 - 12

Choose the bullet for **String**, and then click in the box for **Characters** and type in *1*. Click the **OK** button to return to the **Variable View**. The data type of the variable *bank* is *String* and the **Width** is set to *1*. This data type is referred to as *String1*. Notice that the value in the cell for **Decimals** is grayed out. The graying indicates that this option is not available (strings do not have decimals).

To create **Value Labels** for *bank*, click in the cell for **Values** (the cell currently displays **None**). Click the dots ▦ icon and the **Value Labels** dialog box (Figure 0 - 13) will open.

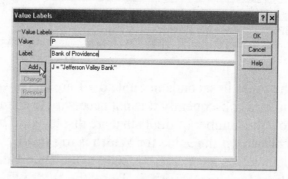

Figure 0 - 13

To create the value label **Jefferson Valley Bank** for **J**, click in the **Value** box and type *J* (this box is case sensitive, that is be sure to type *J* and not *j*), then click in the **Value Label** box and type **Jefferson Valley Bank**, finish by clicking the **Add** button. Create the value label **Bank of Providence** for **P** in the same way. When done click the **OK** button to return to the **Variable View**.

We are now ready to enter the data. Click the **Data View** tab and notice there are now two variables, *minutes* and *bank*, showing in the **Data Editor** with no data. Click in the first row of the variable *minutes* and type 6.5 (the first data value in

Table 0 - 1) and press the enter key or the down arrow key. You are now in the second cell, type *6.6* (the next time), and repeat until all *20* times have been entered.

The order in which the data values for the variable *minutes* are entered does not matter. I entered the *10* Jefferson Valley Bank times followed by the Bank of Providence times. The times could have been entered in any order. For example, I could have alternated Jefferson Valley Bank times and Bank of Providence times (e.g. 6.5, 4.2, 6.6, 5.4, and so on). What is required is that each case of the two variables corresponds to the correct bank. Therefore, the order in which *bank* is entered *must be the same as the order* in which *minutes* was entered. The cases of the variables *minutes* and *bank* are linked.

When you are ready to enter the data values of *bank*, click in the first row of the column for *bank* and type *J* and then press the enter key. Continue in this way until you have entered all the values. If the **View > Value Labels** toggle is set to **Value Labels**, SPSS will display a **down arrow** button ▼ which you can click and select *Jefferson Valley Bank* or *Bank of Providence*. The column width may not be wide enough to display Jefferson Valley Bank or Bank of Providence. If this is the case, go to the **Variable View** and for the variable *bank*, click in the column labeled **Columns** and change the current value to a number greater than 21 (the number of characters in Jefferson Valley Bank).

Section 0-5   Editing and Transforming Variables

If a mistake is found after entering data, it is a simple matter to edit the data. Click in the cell with the incorrect data value, type the new data value, and press the enter key. For example, to change case *4* of the variable *minutes* from *6.8* to *7.0*, click in the cell with *6.8*, type in *7.0,* and press the enter key. This data file may be used later therefore let's change the *7.0* back to *6.8*. To undo the last change made in the **Data Editor**, choose **Edit > Undo Set Cell Value**.

Sometimes it is necessary to create new variables from existing variables in the data file. In statistics, this is referred to as transforming a variable. SPSS provides a way to compute the values for a new variable based on numeric transformations of other variables.

In the Jefferson Valley Bank/Bank of Providence data, we could be interested in computing a new variable that measures the number of seconds that the customer had to wait. Choose **Transform > Compute...** to open the **Compute Variable** dialog box (Figure 0 - 14).

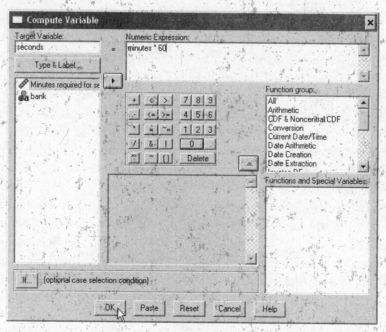

Figure 0 - 14

Since the new variable measures the number of seconds spent waiting, name the variable *seconds* by typing, *seconds* into the **Target Variable** box. Click in the **Numeric Expression** box and type the formula that defines the new variable, *minutes * 60*. When finished, click the **OK** button and the new variable *seconds* will be added to the data file.

The variable name *seconds* is not very descriptive. To create a label for *seconds* click the **Variable View** tab and then click in the cell corresponding to **Label** for the variable *seconds*. In this cell, enter the label *Seconds required for service*. The label could also have been set in the **Compute Variable** dialog box (Figure 0 - 14) by clicking the **Type&Label...** button.

Section 0-6   Printing and Saving a Data File

Sometimes a printed copy of the data file is wanted. For example, a printed copy may be required for a report or other document. In addition, it is easier to check the accuracy of data from a printed copy than from the computer screen. It is always a good idea to double-check the accuracy of recently entered data.

The data file is printed exactly as it appears in the **Data Editor**. If you want to print the data values, make sure the **Data Editor** is in **Data View**. If you want to print the information about the variables then choose the **Variable View** tab prior to printing. To obtain a printed copy of a data file, choose **File > Print...** and the **Print** dialog box will open.

SPSS has two options that can affect the appearance of the **Data Editor**. Choosing **View > Grid Lines** from the menu will hide the grid lines. This is a toggle, which means that choosing **View > Grid Lines** again will make the grid lines reappear. If the data file has value labels, then choosing **View > Value Labels** toggles between displaying data values and displaying value labels.

After typing in a new data file or editing an old data file, the contents of the data editor should be saved. This preserves the data file for future use. To save the contents of the **Data Editor** decide on a name for the data file. The current data refers to the times spent waiting for service at two banks thus the name **Customer waiting time** seems an appropriate choice for this data. To save the contents of the **Data Editor** to the data file named **Customer waiting time**, choose **File > Save As...** and the **Save Data As** dialog box (Figure 0 - 15) will open.

Figure 0 - 15

Enter the name **Customer waiting time** into the **File name** box and click the **Save** button. The data file will be saved as **Customer waiting time.sav** in the selected folder. If you want to save the data file in a different format then click the ⬇ button at the right of the **Save as type** box. A list of available data format types (Microsoft Excel, Lotus 1-2-3, dBase, and text among others) will appear.

Section 0-7   Help in SPSS

SPSS 14.0 for Windows provides several ways to obtain help about how to use SPSS. Every window has a **Help** menu on the menu bar; most dialog boxes have a **Help** button that links directly to a help topic for that dialog box; right-clicking on most objects in a dialog box and selecting **What's This?** will result in a context sensitive description of the item that was clicked on.

Choose **Help > Topics** on the menu bar to open the help window. The help window has tabs for **Contents**, **Index**, **Search**, and **Favorites**. Click the **Contents** tab and double-click the book icons to expand or collapse the contents. Click the **Index** tab to search for a term in the index. Click the **Search** tab to search **Help** for a word or words in Help topics.

Choose **Help > Tutorial** from the menu bar to access the introductory tutorial. The tutorial is displayed using the web browser on your computer. The files were installed on your computer when SPSS was installed and so you do not need an Internet connection to access the tutorial.

Most dialog boxes have a context-sensitive **Help** button. Clicking the **Help** button opens a **Help** window containing information about the current dialog box. Most of the words, buttons, and other items in the dialog box have help information about them that can be accessed by right-clicking on the word, button, or item. If right-clicking results in a little window that says, "**What's This?**" click on "What's This?" and the help will appear.

Choose **Help > Statistics Coach** from the menu bar to get general assistance for many of the commonly used statistical techniques. The **Statistics Coach** asks simple questions, uses nontechnical language, and provides visual examples to help select the basic statistical and charting features that are appropriate for your data. The **Statistics Coach** is not a replacement for taking a statistics course or reading this manual but it can be very helpful at times.

Section 0-8   Exiting SPSS

To exit or quit the SPSS program, choose **File > Exit** from the menu. If you have made changes to the contents of the **Data Editor** since the last time you have saved the data file, SPSS will open a dialog box asking if you want to save the contents of the **Data Editor**. Click the **Yes** button to save the contents of the **Data Editor** prior to SPSS closing; click the **No** button and the contents of the **Data Editor** window will **not** be saved before SPSS closes; and click the **Cancel** button to remain in SPSS (the data file will not be saved and is still available). If the data in the **Data Editor** had already been saved then SPSS closes immediately.

Section 0-9   SPSS Student Version

SPSS Student version is ideal for students who are just beginning to learn statistics. The SPSS Student Version provides affordable, professional statistical analysis and modeling tools to students. The easy-to-learn interface and comprehensive online help system enable students to learn statistics, not software. The SPSS Student

Version has a four-year license code and is for student home use only. It allows up to 50 variables and 1,500 cases. This software is unable to read SPSS command syntax, which is used in a couple of the examples in this manual, and scripting and automation capabilities are unavailable. Add-on modules cannot be used with this package.

Section 0-10   Exercises

1. Data from a Chemistry class experiment designed to measure the **Boiling temperature** of water in degrees Centigrade is displayed in Table 0 - 1. Enter the data into the Data Editor.

Boiling Temperature	Gender of Student	Age of Student	Grade on Experiment
99	M	18	B
103.1	F	22	B
97.4	M	23	C
140	F	19	C
101.2	F	24	A

Table 0 - 1

   a. Name the variables anything you prefer, but their labels must be the headings used above.
   b. Give the variable **Boiling temperature** data type *Numeric5.2*, the variable **Gender of Student** data type *String1*, the variable **Age of Student** data type *Numeric5.0*, and the variable **Grade on Experiment** data type *String2*.
   c. Create value labels for **Gender of Student**. The value label for **M** should be **Male**, and the value label for **F** should be **Female**.
   d. Create a new variable named *error* using the definition

$$\text{Squared Error} = \left(\text{Boiling Temperature} - 100\right)^2 .$$

2. Open the SPSS data file, **Growth study.sav** (this data file comes with SPSS 14.0 for Windows).

   a. How many variables are in this data file?
   b. How many cases are associated with each variable in this data file?
   c. Create a new variable *months*, which is twelve times the variable *age*.

3. The actor, Sean Connery, has played James Bond in seven films. The name of the film and the year the film was released are given Table 0 - 2.

Film	Year
Dr. No	1962
From Russia with Love	1963
Goldfinger	1964
Thunderball	1965
You Only Live Twice	1967
Diamonds are Forever	1971
Never Say Never Again	1983

Table 0 - 2

a. How many cases are there in this data?
b. What would be an appropriate format for the variable *Film*?
c. What would be an appropriate format for the variable *Year*?

4. Data from an Ecology class experiment designed to measure the water temperature of a lake in degrees Fahrenheit is displayed in Table 0 – 4. Enter the data into the Data Editor.

Month	Fahrenheit Temperature
January	37.4
February	33.9
March	41.6
April	50.8
May	53.0
June	

Table 0 - 3

a. How many variables are in this data file?
b. How many cases are there in this data?
c. Create a new variable named *centigrade* using the definition

$$C = \frac{5}{9}(F = 32).$$

Chapter 1

Introduction to Statistics

Chapter 1   Introduction to Statistics

You should be familiar with SPSS before beginning this chapter. Specifically you should be able to create a new data file or open an existing data file, edit and transform data values in the data file, save and print the data file. In this chapter, we learn how to obtain simple random samples and systematic random samples from a population using SPSS; consult *Elementary Statistics*, 10/e or *Essentials of Statistics*, 3/e for a full discussion of the topics.

Section 1-1   Data Types

SPSS is a computer program and cannot make certain judgments about data. For SPSS to make accurate calculations, the level of measurement (nominal, ordinal, interval, or ratio) must be set for the variables. Variables that have **Nominal** measurement level represent categories with no built-in ordering (e.g. male, female). Variables that have **Ordinal** measurement level represent categories with some built-in ordering (e.g. low, medium, high or strongly agree, agree, disagree, strongly disagree). SPSS does not differentiate between interval and ratio scale; both data types are classified as **Scale** measurement level. A variable with **Scale** measurement level must be numeric with the property that the difference between two values has meaning (e.g. age, income, temperature).

SPSS makes an educated guess about the appropriate data type when the variable is created but you should always check the data type. SPSS usually classifies qualitative (nonnumerical) variables as **Ordinal** and quantitative (numerical) variables as **Scale**.

Open the **Employee data** file. This data file comes with SPSS 14.0 for Windows. If you do not know how to open this data file refer to Section 0-2 of this manual. Choose the **Variable View** tab in the Data Editor to display information about the variables. The right-most column in the Variable View labeled **Measure** (see Figure 1 - 1) shows the current measurement level. For example, *id* has measurement level **Scale,** *gender* has measurement level **Nominal,** *bdate* has measurement level **Scale**, and *educ* has measurement level **Ordinal**, and so on.

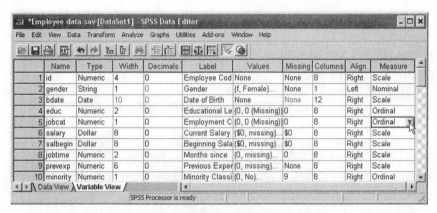

Figure 1 - 1

Since there is no inherent ordering to the values of the variables, *jobcat* and *minority*, their measurement level should be **Nominal**. For these two variables, it appears that SPSS made an incorrect assignment based on its educated guess. To change the measurement level, click in the cell that you want to change and a **down arrow** button ▼ will appear. Click the down arrow and a drop-down menu of available options for the measurement level appears. Choose **Nominal** from the list to change the measurement level of *jobcat* to **Nominal**. Change the measurement level for *minority* to **Nominal** in the same way.

Section 1-2   Selecting a Simple Random Sample

The study of statistics is concerned with summarizing, analyzing, and drawing conclusions from data to answer questions. Sometimes all the cases of interest in a particular problem will be studied. The collection of all the possible cases (subjects, scores, people, or measurements) is called the **population**. Populations may be too large to study or may be too expensive or time consuming to measure all the cases in the population. Therefore, sometimes only a subcollection (subset or part) of the population is studied; we call this subcollection of the population a **sample** from the population.

Data are summarized or described by calculating numbers that describe certain characteristics (for example, the average, minimum, or median) of the data. Numbers that describe characteristics of a population are called **parameters**. Numbers that describe characteristics of a sample are called **statistics**. A major goal of studying Statistics is to learn how to describe *population parameters* by *sample statistics*. In order to do this, the samples need to be chosen in an appropriate way.

In Statistics, we study **samples** *chosen randomly* from a population. There are many different ways to choose random samples from a population; the most common is called a **Simple Random Sample**. A **Simple Random Sample** (SRS) is a sample of *n* cases selected in such a way that every possible sample of the same size *n* has the same chance of being chosen.

Open the **Employee data** file if it is not already open. To select a simple random sample of size *30* from the population of the *474* employees in the **Employee data**, choose **Data > Select Cases…** from the menu and the **Select Cases** dialog box (Figure 1 - 2) will open.

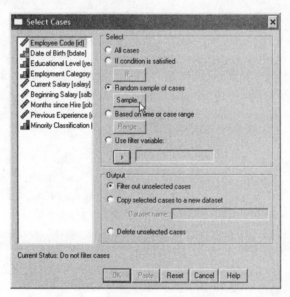

Figure 1 - 2

From the **Select Cases** dialog box, choose the bullet for **Random sample of cases** and then click the **Sample...** button to open the **Select Cases: Random Sample** dialog box (Figure 1 - 3).

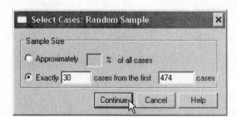

Figure 1 - 3

Choose the bullet for **Exactly** to select a **Simple Random Sample**. The two boxes on the line beginning with **Exactly** are no longer grayed out and are now ready to accept information. Click inside the first box and type *30* and then click inside the second box and type *474* (recall an SRS of size *30* from the *474* cases is required). Once these values have been entered, click the **Continue** button, which will close the **Select Cases: Random Sample** dialog box and return to the **Select Cases** dialog box (Figure 1 - 2). The **Select Cases** dialog box now reflects that *30* cases from the *474* will be sampled.

Before selecting the random sample, a decision regarding how to handle the unselected cases should be made. There are three options in Figure 1 - 2. The first option, **Filtered,** instructs SPSS to keep all the cases as part of the data file but only use the selected cases (the *30* values in the SRS) when it does calculations. This choice is useful when the entire population might be needed for later analysis. The second option, **Copy selected cases to a new data set,** copies only the selected cases to a different data set having a specified name. The third option, **Deleted,** instructs SPSS to delete the unselected cases so that the current data file will only include the *30* cases. This choice is useful when the random sample will be printed and the population will not be needed later. Select **Filtered,** this is the safest choice and the default, and then click the **OK** button.

The **Data View** (Figure 1 - 4) has changed. First, some of the cases have a slash through their case number. The slash indicates that the case was not selected as part of the SRS. In addition, SPSS has created a new variable called, *filter_$*. This new variable takes on only two values, zero and one. Zero indicates that the case was not selected and one indicates that the case was selected. If **View > Value Labels** has been toggled, then the **Data View** will show the value labels "Not Selected" and "Selected" instead of zero and one.

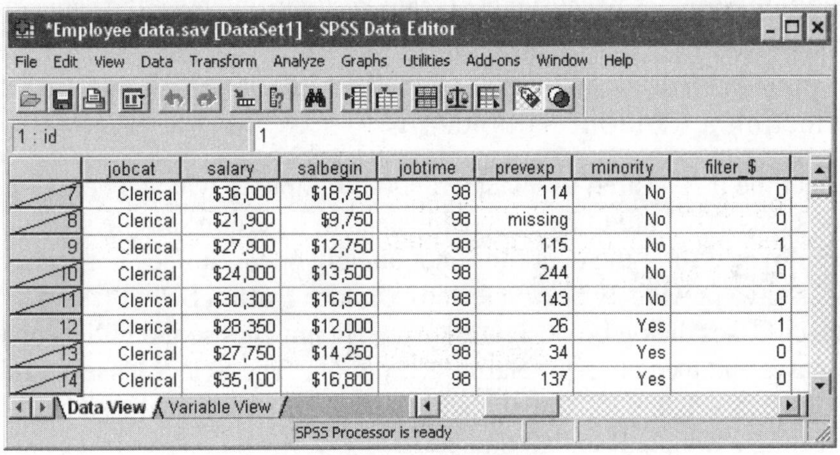

Figure 1 - 4

Do not be concerned if the selected cases in your Data View are different from those shown in Figure 1 - 4. The SRS chosen by SPSS is only one of many possible random samples that could result, there are about 1.4×10^{20} different simple random samples of size *30* that can be chosen from a population of *474* cases. If this process for obtaining a simple random sample is repeated, the new simple random sample will be different from the previous random sample that was obtained. SPSS will not create a new filter variable for the second and succeeding samples, it only changes (updates) the variable *filter_$* and the case numbers with slashes. Although the random samples that are chosen are different each time, it is a cornerstone of statistics that each of these random samples consistently and accurately describes the population.

If at some point, you need to do calculations on all the data, then choose the bullet for **All Cases** in the **Select Cases** dialog box (Figure 1 - 2) to turn off the filtering. This will instruct SPSS to use all the cases when calculating statistics or doing analyses. The variable *filter_$* will still exist and can be used to return to the random sample so no information is lost.

Section 1-3   More Random Samples

The previous section discussed selecting a **Simple Random Sample**. There are many ways to select a sample randomly from a population. For example, the random sample could be selected in such a way that each case had an equal chance of being selected. Your textbook calls this a **Random Sample**. A Simple Random Sample differs from a Random Sample in that the sample size, *n* for a SRS is chosen in advance. Another common sample is called a **Systematic Sample**. Systematic samples select one of the first *k* cases at random and then every k^{th} case after that is selected. SPSS can choose samples according to both these sampling methods as well as many others. **Random Samples** and **Systematic Samples** are not **Simple Random Samples** because not all samples of size *n* have the same chance of being selected.

Selecting a Random Sample

Sometime it is desired to select a random sample in which a certain percentage of the cases (say *10%*) are to be selected instead of a certain sample size. Open the **Employee data** if the data file is not already open. It does not matter if some of the cases have previously been selected. Choose **Data > Select Cases...** to open the **Select Cases** dialog box (Figure 1 - 2). Again, choose the bullet for **Random sample of cases** and then click the **Sample...** button. The **Select Cases: Random Sample** dialog box (Figure 1 - 3) will open. To this point, this is the same process for selecting a Simple Random Sample.

To select Random Sample of *10%* of the cases, choose the bullet for **Approximately** and type *10* into the box. Once this is done, click the **Continue** button. SPSS will return to the **Select Cases** dialog box (Figure 1 - 2), which now reports "Approximately *10%* of cases" will be selected.

Before clicking the **OK** button, decide what to do with unselected cases. It is easier to see the selected sample if the unselected cases are deleted. Choose **Deleted** and click the **OK** button to select the random sample. **Do not save** the current data file to **Employee data** otherwise the original file will be *permanently* replaced by this sample. Choose **File > Save As...** to save the contents of the Data Editor to the data file **Random Sample** (see Section 0-6 for information about saving data files).

The data file **Random Sample** will have about *10%* of the number of cases as the population. Since the population has *474* cases then this random sample will have about *47* cases. The number of cases in the random sample will vary because SPSS selects each case independently with probability *10%*. The percentage of cases selected is therefore only approximate. For this data, random samples having between *34* and *60* cases are common.

Notice that the first case of the data file **Random Sample** (Figure 1 - 5) was the ninth case in the original **Employee data** (Figure 0 - 6 from Section 0-2 of this manual). Because we chose to delete unselected cases, only the selected cases are a part of the data file. Notice none of the cases in the **Random Sample** data file has a slash through their case number.

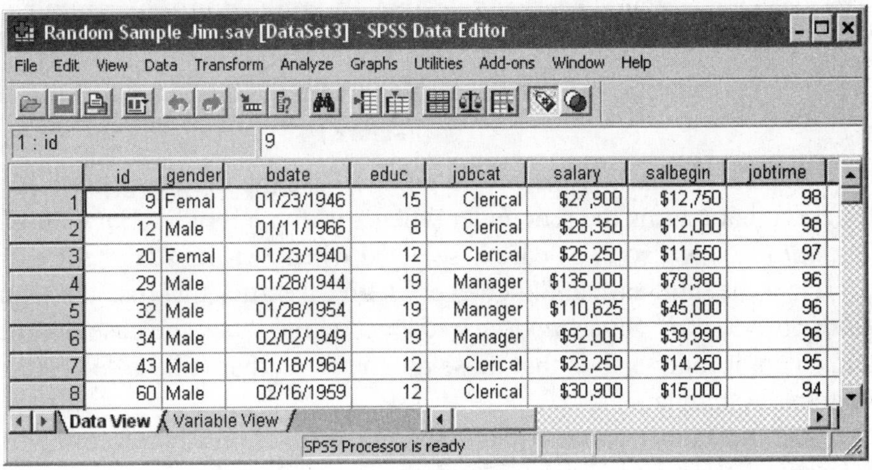

	id	gender	bdate	educ	jobcat	salary	salbegin	jobtime
1	9	Femal	01/23/1946	15	Clerical	$27,900	$12,750	98
2	12	Male	01/11/1966	8	Clerical	$28,350	$12,000	98
3	20	Femal	01/23/1940	12	Clerical	$26,250	$11,550	97
4	29	Male	01/28/1944	19	Manager	$135,000	$79,980	96
5	32	Male	01/28/1954	19	Manager	$110,625	$45,000	96
6	34	Male	02/02/1949	19	Manager	$92,000	$39,990	96
7	43	Male	01/18/1964	12	Clerical	$23,250	$14,250	95
8	60	Male	02/16/1959	12	Clerical	$30,900	$15,000	94

Figure 1 - 5

Selecting a Systematic Sample

A **Systematic Sample** is a sample in which the first case is selected at random from the first *k* cases, then every k^{th} case after that is selected. We will describe how to select a systematic sample of about *30* data values. Since there are *474* data values in the **Employee data** file it follows, we should choose *k* to be about *474 / 30*, or about *15*. Therefore, we will select systematic sample by choosing every 15th case from the **Employee data** file.

Open the **Employee data** if it is not already open. It does not matter if some of the cases have already been selected. Choose **Data > Select Cases...** to open the **Select Cases** (Figure 1 - 2) dialog box. Choose the bullet for **If condition is satisfied** and click the **If...** button; the **Select Cases: If** dialog box (Figure 1 - 6) will open.

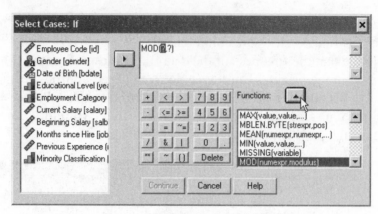

Figure 1 - 6

To select every k^{th} case we are going to use the **mod** function. *Mod(x, y)* is a function that returns the remainder when *x* is divided by *y*. For example, *mod(5, 3)* is *2*; *mod(32, 3)* is also *2*. A case is selected whenever the statement in the box is true. To select the third case and then every fifth case after that, use the expression *mod(case, 5) = 3*. Since *mod(case, 5) = 3* is true when the case number is *3, 8, 13*, and so on, it follows then that those are the cases that will be selected for the systematic sample.

The **mod** function is far down the list of functions therefore it will be necessary to use the ▼ button to browse down the list. Select the **mod** function by clicking on **MOD(numexpr, modulus)** and then press the **paste expression** ▲ button to paste MOD(?, ?) into the expression box.

The first question mark (numexpr) should be a counter for the case; the variable *id* can be used as a substitute for case. The first question mark is highlighted automatically. Click on **Employee Code [id]** to select the variable *id* and then click the **variable paste** ▶ button, which will replace the question mark with *id*.

The second question mark (modulus) is the value of *k*. Recall, we are selecting a systematic sample by choosing every *15*th case from the **Employee data** file. The first case will be selected at random from the first *15* cases and then every *15*th case after that will be selected. Therefore, a modulus of *15* should be entered.

A random number table can be used to choose the first sample. The first sample can also be computed in SPSS using the **Compute Variable** dialog box (Figure 0 - 14 in Section 0-5 of this manual). For example, typing the expression RND(UNIFORM(15) + 0.5) into the **Numeric Expression** box will compute random integers between *1* and *15*. Uniform is a function that returns random values between *0* and *15*. These values are not integers and so we need to use the RND function to round the numbers to the nearest integer. We add one-half prior to rounding so that the values *0.5-1.5* round to *1*, *1.5-2.5* round to *2*, etcetera. In this case, it is probably easier to simply choose the first sample by selecting a value from a random number table.

Suppose the result of choosing a number at random between *1* and *15*, inclusive, results in the value of *6*. Edit the expression MOD(id, ?) so that it reads MOD(id, 15)

= 6 (see Figure 1 - 7). Alternately, the expression MOD(id, 15) = 6 could have been typed into the box. When you are finished, the **Select Cases: If** dialog box should look like Figure 1 - 7.

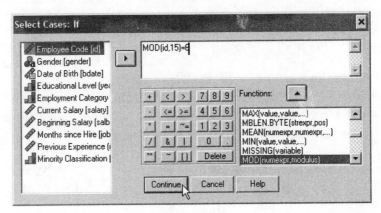

Figure 1 - 7

Click the **Continue** button and the **Select Cases: If** dialog box will close and return to the **Select Cases** dialog box (Figure 1 - 2). The **Select Cases** dialog box now displays Mod(id,15)=6 to the right of the **If…** button, to reflect how the cases will be selected from the data file.

Choose the bullet for **Filtered** to filter unselected cases and then click the **OK** button. The Data View (Figure 1 - 8) now shows that the sixth case is selected and every *15*th case after that (i.e. the *6*th, *21*st, *36*th…).

	jobcat	salary	salbegin	jobtime	prevexp	minority	filter_$
1	Manager	$57,000	$27,000	98	144	No	Not Selected
2	Clerical	$40,200	$18,750	98	36	No	Not Selected
3	Clerical	$21,450	$12,000	98	381	No	Not Selected
4	Clerical	$21,900	$13,200	98	190	No	Not Selected
5	Clerical	$45,000	$21,000	98	138	No	Not Selected
6	Clerical	$32,100	$13,500	98	67	No	Selected
7	Clerical	$36,000	$18,750	98	114	No	Not Selected
8	Clerical	$21,900	$9,750	98	missing	No	Not Selected
9	Clerical	$27,900	$12,750	98	115	No	Not Selected

Figure 1 - 8

Section 1-4   Exercises

1. Open the SPSS data file, **Breast cancer survival** (this data file comes with SPSS 14.0 for Windows). This data file has *1207* cases and *11* variables

    a. What are the value labels associated with the two variables, *pathsize* and *er*?
    b. What are the measurement levels associated with the variables *age*, *lnpos*, and *Pathological Tumor Size (Categories)*?
    c. Should any of the measurement levels be changed to more accurately show their true nature? If so, which one(s) should be changed and to which measurement value should they be changed?

2. Select a **Simple Random Sample** of size *10* from the **Breast cancer survival** data file. Save the sample to a data file named, **SRS**. Print the *10* values in this simple random sample.

3. Select a **Systematic Sample** from the **Breast cancer survival** data file by choosing the fourth case and then every *100*th case. Print the *13* data values in this systematic sample.

4. Select a **Systematic Sample** from the **University of Florida Graduate Salaries** data file. This data file comes with SPSS 14.0 for Windows as **University of Florida graduate salaries.sav**. Choose the fifth case and then every *150*th case. Print the *8* data values in this systematic sample.

Chapter 2

Summarizing and Graphing Data

Chapter 2   Summarizing and Graphing Data

There are many tools used in Statistics to visualize, summarize, and describe data. This chapter will focus on using SPSS to create the tables, graphs, and charts, and to calculate the descriptive statistics that are discussed in your textbook. Consult Chapter 2 of *Elementary Statistics*, 10/e or *Essentials of Statistics*, 2/e for a full discussion of the topics.

Section 2-1   Frequency Distributions

For a first look at data, it is practical to begin by describing the distribution of values in a data file. Many statistical tools have been developed for this purpose. SPSS can simplify the creation of the tables, graphs, and charts used for exploring a data file. A common tool used for describing the distribution of values is the **frequency distribution**.

SPSS makes a **frequencies report**. A frequencies report is different from a frequency distribution. The frequencies report simply lists distinct data values with their frequencies; it does not list the frequency of data values in a list of categories. SPSS does not provide a way to customize the class limits to enable a frequency table to be created. It is not difficult to make a frequency distribution by hand from the frequencies report.

Frequencies Report for a Variable

The data listed in Data Set 1: **Health Exam Results** in Appendix B of your textbook (this data is saved on the data disk as **Mhealth.sav**) has *13* different measurements on *40* men. The variables being measured are age, height, weight, waist circumference, pulse rate, systolic and diastolic blood pressure, cholesterol level, body mass index, upper leg length, elbow breadth, wrist breadth, and arm circumference. These data are from the U.S. Department of Health and Human Services, National Center for Health Statistics, Third National Health and Nutrition Survey.

Open the **Mhealth.sav** data file (see Section 0-3). To obtain a frequencies report for *age* and *ht*, choose **Analyze > Descriptive Statistics > Frequencies…** to open the **Frequencies** dialog box (Figure 2 - 1).

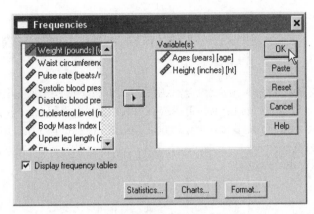

Figure 2 - 1

Highlight the variable *age* by clicking on its label "**Ages (years) [age]**" and then click the **variable paste** ▶ button to copy the variable to the **Variable(s)** list. In a similar manner, copy the variable *ht* to the **Variable(s)** list.

If the variable has many distinct data values the resulting frequencies report can be very long. The frequencies report(s) can be suppressed by unchecking the checkbox for **Display frequency tables** in the **Frequencies** dialog box. If you uncheck this box, SPSS may give a warning that says, "You have turned off all output. Unless you request Display Frequency Tables, Statistics, or Charts, FREQUENCIES will generate no output." Frequency reports are often the first tool used to describe a data file. SPSS provides buttons on the **Frequencies** dialog box for calculating some common statistics and charts that are useful for describing data files. In the following sections, we will discuss many of these charts and statistics. The checkbox for **Display frequency tables** should be checked now since the goal here is to produce the frequencies table.

Click the **OK** button and SPSS will produce the frequencies report in a new window called the **Output Viewer** window**.** The frequencies reports for both *age* and *ht* are rather long and so only the beginning and end of the frequencies report for *age* is shown below (Figure 2 - 2). The Output Viewer window displays the entire frequencies report for both variables.

Ages (years)

		Frequency	Percent	Valid Percent	Cumulative Percent
Valid	17	2	5.0	5.0	5.0
	18	2	5.0	5.0	10.0
	20	4	10.0	10.0	20.0
	22	1	2.5	2.5	22.5
	25	1	2.5	2.5	25.0
	54	1	2.5	2.5	90.0
	55	1	2.5	2.5	92.5
	56	1	2.5	2.5	95.0
	58	1	2.5	2.5	97.5
	73	1	2.5	2.5	100.0
	Total	40	100.0	100.0	

Figure 2 - 2

The frequencies report shows the frequencies associated with each distinct data value. Be careful when reading a frequencies report, for example, a common oversight is to not notice that there are no data values equal to *19, 21, 23, 24,* and so on in this data file. The frequencies report indicates that the minimum and maximum data values are *17* and *73,* respectively. The report shows that the data value *17* occurs two times, and that the data value *20* occurs four times. The last line of the frequencies report indicates that there are *40* data values in the data file. A frequency distribution can easily be obtained from the frequencies report. The column labeled, Percent, indicates the relative frequency with which each data value occurs. This column can be used to help make a relative frequency distribution of the ages. The last column labeled, Cumulative Percent, gives the cumulative relative frequencies. This column can be used to make a cumulative frequency distribution for ages.

Output Viewer Window

This section provides an overview of the Output Viewer window. The Output Viewer window shows the output that results from a command procedure being run in a dialog box. If there is no Output Viewer window open, SPSS will open a new Output Viewer window. The Output Viewer window can be used to hide, re-center, or delete the results of procedures. The frequencies procedure results are shown in the Output Viewer window (Figure 2 - 3).

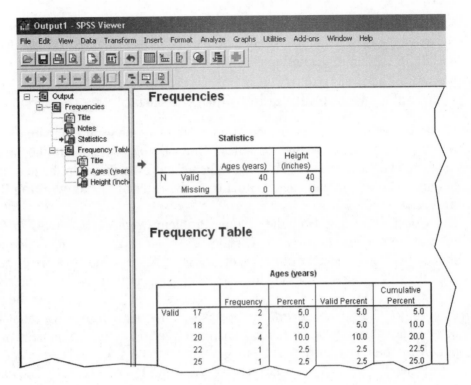

Figure 2 - 3

   The SPSS Output Viewer window is divided into two independent windows. The left hand window (**Outline window**) shows the output in outline view. The right hand window (**Output window**) shows the output.

   The Outline window is useful for choosing output to be displayed in the Output window. Individual items in the Outline window can be displayed or hidden. For example, **Notes,** which has information about the execution of a procedure, is hidden by default. Click on **Notes** (or the icon 🔲 in front of it) to select the item **Notes** and choose **View > Show** to cause the Notes to be displayed in the Output window. Choose **View > Hide** to hide a selected item in the Output window.

   Clicking the minus sign, ⊟ in front of an item will hide all the output associated with the item. This is the same as selecting the item and choosing **View > Collapse**. Click on the **minus sign** in front of **Frequencies**; notice that the Output window is now blank. Click on **Frequencies** to select it, and then choose **View > Expand** to make the output reappear. Alternately, clicking on the **plus sign** ⊞ in front of a collapsed item will also expand the output.

   Clicking on (selecting) an item in the Outline window will cause the Output window to be re-centered on the item. For example, click on **Ages (years)** in the Outline window to display the frequencies report for the variable *age* in the Output window. Click on **Height (inches)** in the Outline window and the display in the Output window changes to the frequencies report for the variable *ht*.

   If you want to delete an item from the Output window, then click on the item and choose **Edit > Delete**. For example, to delete the frequencies report for height, click on **Height (inches)** in the Outline window and choose **Edit > Delete** (or simply press the delete key). If you mistakenly delete some output you may choose **Edit > Undo** and SPSS will undo the previous action.

Frequencies Reports for subgroups of a Variable

Often the same variable is measured for more than one group. In situations like this, the data file will contain two variables. One variable will contain the values being measured and a second variable that indicates into which group the particular case belongs. An example of such a situation is the **Ages of Oscar-winning Best Actors and Actresses**.

Consider the data from the Chapter Problem *Do the Academy Awards involve discrimination based on age?* in Chapter 2 of your textbook. The table on page 41 of the textbook lists the ages of recipients for Best Actress and Best Actor beginning with the first awards ceremony in 1928. The data contains two variables, *age* and *gender*. The variable *age* represents the age of the actor or actress at the time of the ceremony. The variable *gender* indicates whether the recipient was an actress or actor. The table lists the ages for 76 actresses and 76 actors. You need to create a data file called **Academy Awards Ages** having the two variables *age* and *gender*. Create labels for *gender* so that **M = "Actor"** and **F = "Actress"**. The data file will have 152 cases, 76 ages for actresses who were recipients of the award and 76 ages for actors who were recipients. The **Variable View** (Figure 2 - 4) of the data file and the **Data View** (Figure 2 - 5) are shown below.

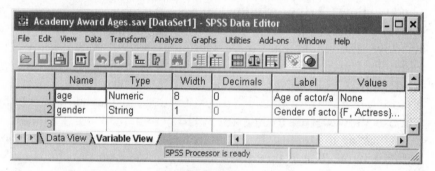

Figure 2 - 4

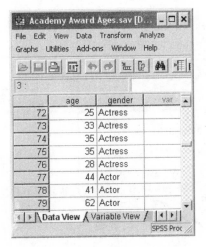

Figure 2 - 5

To obtain the frequencies reports of the ages for the actress and actor groups, choose **Analyze > Descriptive Statistics > Crosstabs...** to open the **Crosstabs** dialog box (Figure 2 - 6). The **Crosstabs** procedure forms two-way (and even three-way) tables. **Crosstabs**, like the **Frequencies** procedure, creates a frequencies report and calculates a variety of statistics and graphical displays.

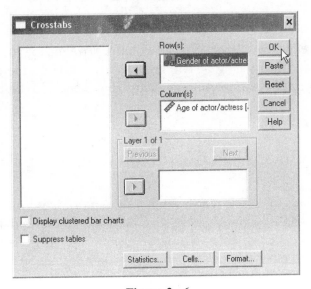

Figure 2 - 6

Copy the variable *age* to the **Row(s)** list and the variable *gender* to the **Column(s)** list by selecting the variable and clicking the **variable paste** button. Click the **OK** button and the frequencies report will appear in the Output Viewer window. The frequencies reports are very long; only the beginning and end of the frequency report is shown in Figure 2 - 7.

Age of actor/actress * Gender of actor/actress Crosstabulation

Count

		Gender of actor/actress		Total
		Actress	Actor	
Age of actor/actress	21	1	0	1
	22	1	0	1
	24	2	0	2
	25	4	0	4
	26	4	0	4
	60	1	2	3
	61	1	0	1
	62	0	3	3
	63	1	0	1
	74	1	0	1
	76	0	1	1
	80	1	0	1
Total		76	76	152

Figure 2 - 7

The frequencies report gives the number (frequency) of data values for a particular age for each of the two groups. For example, there were four actresses age 25 who received the Best Actress award, and there were three actors age 62 who won the award.

Section 2-2   Graphs and Charts for Visualizing Data

It is said that a picture is worth a thousand words. Describing the distribution of data can be accomplished much more succinctly with pictures than with a frequencies report or a frequency table. Histograms, pie charts, bar charts, stem-and-leaf plots, boxplots, and other graphs are pictures of the information in a frequency table. A **histogram** is a picture of the information in the frequency table that shows the shape, center, and spread of the distribution. When the data have nominal or ordinal levels of measurement, **pie charts** and **bar charts** can be used to describe the data. **Stem-and-leaf plots** are useful when the number of data values is not too large (say less than *100*). They provide a way to see the shape of the distribution without losing the original data values. **Boxplots** show the shape of the distribution and also provide a way to check for outliers in the data file.

The **Frequencies** procedure can be used to create graphical displays and calculate descriptive statistics for a single factor. To create bar charts, pie charts, or histograms open the **Frequencies** dialog box (Figure 2 - 1) and click the **Charts...** button. We will discuss using the **Explore** procedure to produce descriptive statistics and graphical displays.

Histograms

To make histograms of the ages of the actress and actor groups, choose **Analyze > Descriptive Statistics > Explore...** to open the **Explore** dialog box (Figure 2 - 8).

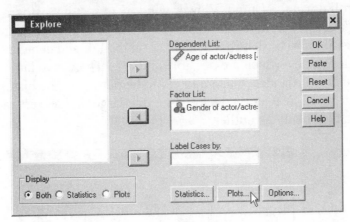

Figure 2 - 8

Copy the variable *age* to the **Dependent List** and the variable *gender* to the **Factor List** by selecting the variable and clicking the **variable paste** button. Next, click the **Plots...** button to open the **Explore: Plots** dialog box (Figure 2 - 9).

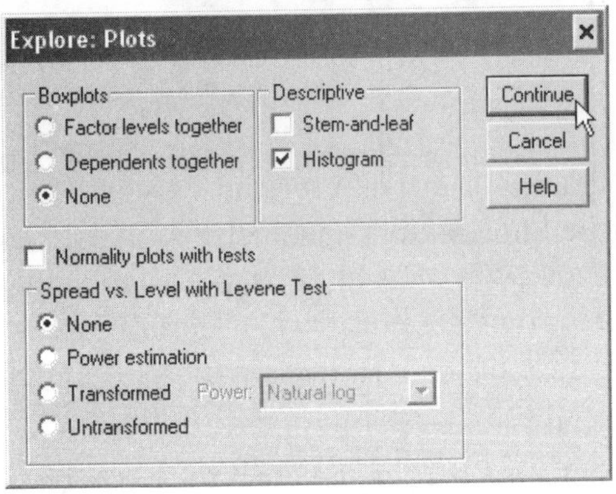

Figure 2 - 9

Notice that Figure 2 - 9 has both bullets and checkboxes. Checkboxes and bullets serve two different purposes. Within a grouping, only a single bullet may be chosen while multiple checkboxes can be selected. For example, if the checkboxes for **Histogram** and **Stem-and-leaf** are both chosen then both a histogram and stem-and-leaf plot will be produced.

Choose the checkbox for **Histograms** to make a Histogram for each combination of the factor levels (each group) of the variable in the Factor List and variable in the Dependent list. For example, in this problem histograms for actresses and actors of the *gender* variable will be produced. This same dialog box can be used to make **Stem-and-leaf Plots** and **Boxplots**, as well as some other plots that we will not discuss here. For now, choose the bullet for **None** under Boxplots. Click the **Continue** button to return to the **Explore** dialog box and then click the **OK** button to display the histograms in the Output Viewer window.

Scroll down the Output Viewer window to see the histograms for actresses and actors. The histogram for the actresses group (Figure 2 - 10) shows the distribution of the data values to be distributed heavier on the left and centered around 35, except for five larger values, which are possible outliers. These five large values are separated from the rest of the data and are likely to be outliers. The histogram for the actors is also distributed heavier on the left with one larger value, separated from the rest of the data, also likely an outlier.

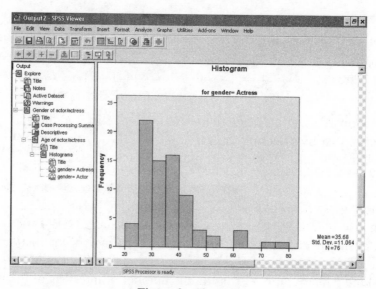

Figure 2 - 10

Chart Editor

Histograms, in fact any chart created in SPSS, can be customized. To customize the histogram for *actresses*, first select the histogram by clicking on **gender = Actress** in the Outline window and then choose **Edit > SPSS Chart Object > Open** to open the chart editor (Figure 2 - 11). Alternately, you can double-click on the Histogram to open the Chart Editor. The **Chart Editor** window has its own menu (notice the menu along the top of the Chart Editor window has changed).

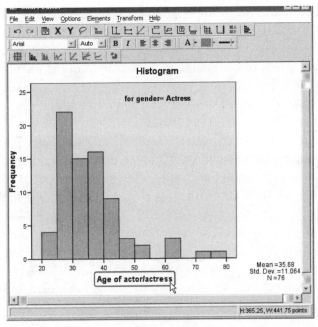

Figure 2 - 11

There are two axes, the **Interval Axis** (the axis from which the bars originate) and the **Scale Axis** (the axis which displays numerical values to scale). To modify the **Interval Axis** Title, click on the title once. The text has a blue circle around it (Figure 2 - 11). Click on the text once again to start edit mode where you can change the title. Change the title to **"Age of Actress Receiving Award."** Then click anywhere else on the screen to view the results. The title can also be justified left, centered, or right by clicking on the toolbar. The label **"Frequency"** can be modified in a similar manner.

Click on any of the **Tick marks** on either axis and then choose **Edit > Properties.** The **Properties Dialog Box** will open (Figure 2 –12). This dialog box can also be used to show or hide **Tick marks** and **Grid Lines**. These modifications are cosmetic and will have no affect on the shape of the histogram.

SPSS automatically determines the number of intervals and interval widths when making a histogram. Changing the number of intervals or the interval width will have an affect on the shape of the histogram. Click on any of the ages along the **Interval Axis** and then choose **Edit > Properties** and then click on the **Scale** tab (Figure 2 - 12). Change the **major increment** from 10 to 5 and click **Apply**.

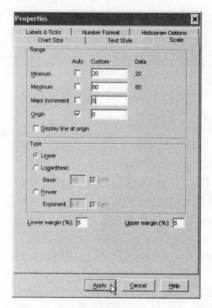

Figure 2 - 12

This will create classes of *20-25, 25-30, 30-35, 35-40, 40-45,* and so on. This notation might be confusing because it is not be clear into which class a data value of *25* would be included. SPSS understands the interval *20-25* to be $20 \leq x < 25$, therefore *25* goes into the interval labeled *25-30*.

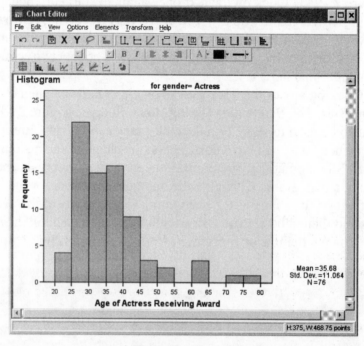

Figure 2 - 13

The histogram has now been updated; you can now make more changes or close the Chart Editor. Choose **File > Close** from the menu to close the Chart Editor.

The shape of the histogram for actresses (Figure 2 - 13) and actors (not shown) indicates that the ages fall mainly into the *20's* through the *50's*. Further analysis will be necessary to determine if the larger data values are outliers.

Stem-and-leaf plots

To make **stem-and-leaf plots** for the ages for the two groups: Actresses and Actors, choose **Analyze > Descriptive Statistics > Explore…** to open the **Explore** dialog box (Figure 2 - 8). Click the **Plots…** button to open the **Explore: Plots** dialog box (Figure 2 - 9). Choose the checkbox for **Stem-and-leaf** and click the **Continue** button to close the dialog box. Click the **OK** button in the **Explore** dialog box and the stem-and-leaf plots will appear in the Output Viewer window (Figure 2 - 14).

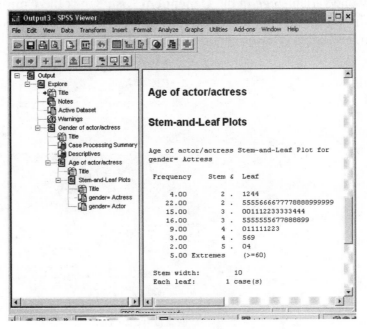

Figure 2 - 14

SPSS only shows the leading digit and the next digit in the number in the stem-and-leaf plot; that is, it truncates the data values to only two digits. To know the size of the data values you must read the Stem width. In the plot above, the Stem width is *10*, which means that the first two stems in the plot display data values between *20-24* and 25-29, respectively. The third and fourth stems in the plot display data values between *30-34* and *35-39*, and so on. Since the age data values are no more than two digits, it is possible to determine the exact values directly from the stem-and-leaf plot. On the first 2-stem, the data values can be determined by looking at the leaves. The data values are *21, 22,* and two *24's*. On the second 2-stem, the data values are four *25's*, four *26's*, four *27's*, four *28's*, and six *29's*.

The stem-and-leaf plot helps us to represent a body of data in a comprehensive way, and gives a feel for the "shape" of the distribution similar to that of the histogram. The stem-and-leaf plot indicates that there are five extreme cases which we need to investigate as outliers.

Boxplots

To make boxplots for the ages for the two groups: Actresses and Actors, choose **Analyze > Descriptive Statistics > Explore…** to open the **Explore** dialog box (Figure 2 - 8). Click the **Plots…** button to open the **Explore: Plots** dialog box (Figure 2 - 9). Choose the bullet for **Factor levels together** to make a separate Boxplot for each of the variables in the **Dependent List** in the **Explore** dialog box. If there are several variables in the **Dependent List**, choose the bullet for **Dependents together** to obtain side-by-side Boxplots. In this case, it does not matter since there is only one variable *age* in the **Dependent List**. Click the **Continue** button, and then click the **OK** button in the **Explore** dialog box, and the boxplots will appear side-by-side in the Output Viewer window (Figure 2 - 15).

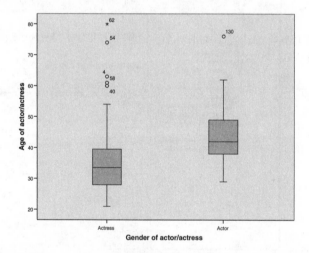

Figure 2 - 15

The boxplots clearly show that the ages for actors are a little older than the ages for the actresses. The ages for the actors are skewed to the right. There are five outliers for the actress group and only one outlier for the actor group.

Section 2-3   Scatter Diagrams

A **Scatter Diagram** (also known as a **Scatterplot**) is a plot of paired (x, y) data with a horizontal x-axis and a vertical y-axis. The y-axis variable determines the vertical position of the point and the x-axis variable determines the horizontal position of the point. Scatter diagrams are useful for exploring the relationship between two variables. Exploring the relationship between two variables will be discussed in more detail in Chapter 10.

The data listed in Data Set 1: **Health Exam Results** in Appendix B of your textbook (this data is saved on the data disk as **Mhealth.sav**) includes the weight (in pounds) and the waist circumference (in cm) for 40 males. Make a scatter diagram to determine if there is a relationship between the weight and waist circumference measurements.

Open the **Mhealth** (see Section 0-3) data file. Make a **Scatter diagram** of weight versus waist circumference by choosing **Graphs > Scatter/Dot...** to open the **Scatterplot** dialog box (Figure 2 - 16).

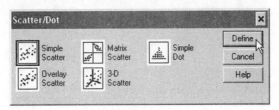

Figure 2 - 16

Choose the icon for **Simple** and then click the **Define** button to open the **Simple Scatterplot** dialog box (Figure 2 - 17).

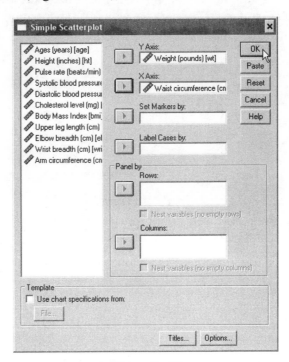

Figure 2 - 17

Select the **Weight (pounds) [wt]** for the Y Axis and **Waist circumference (cm) [waist]** for the X Axis by clicking on the variable label and then clicking the **variable**

paste button. Click the **Titles...** button if you want to give your Scatter diagram a title. Click the **OK** button and the Scatterplot (Figure 2 - 18) will appear in the Output Viewer window.

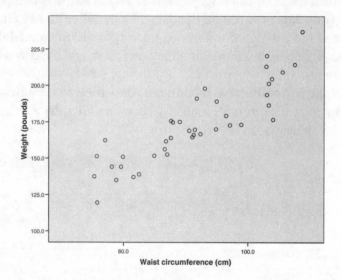

Figure 2 - 18

The scatter diagram indicates that as the waist circumference increases the weight of the person tends to increase as well. There appears to be an approximately linear relationship between the weight and the waist measurements.

Section 2-4   Exercises

1. Refer to Data Set 12: **Weights and Volumes of Cola** in Appendix B (this data is saved on the data disk as **Cola.sav**). Construct a histogram for the weights of Regular Cola by starting the first class at *0.7900* lb. and use a class width of *0.0050* lb. Then construct another histogram for weights of Diet Cola by starting the first class at *0.7500* lb and use a class width of *0.0050* lb. Then compare the results and determine whether there appears to be a significant difference.

2. Refer to Data Set 6: **Bears (wild bears anesthetized)** in Appendix B (this data is saved on the data disk as **Bears.sav**). Construct a histogram for the measured weight of the bears with *11* classes with a lower class limit of *0* and a class width of *50* lbs.

3. Refer to Data Set 12: **Weights and Volumes of Cola** in Appendix B (this data is saved on the data disk as **Cola.sav**). The data set contains the weight and volume measurements on 36 cans of Regular Cola, Diet Cola, Regular Pepsi, and Diet Pepsi. Make side-by-side boxplots of the weights of Regular Cola, Diet Cola, Regular Pepsi, and Diet Pepsi. What can you conclude about the distributions of the Colas?

4. The data listed in Data Set 9, **Electric Consumption of a Home** (this data is saved on the data disk as **ELECTRIC.sav**) are average daily *temperatures* and the corresponding amounts of *energy consumption (kWh)*. Make a Scatter diagram of consumption versus temperature for the first *10* cases. Based on your results, is there a relationship between average daily temperatures and energy consumed?

Chapter 3

Statistics for Describing, Exploring, and Comparing Data

Chapter 3   Statistics for Describing, Exploring, and Comparing Data

This chapter, like Chapter 2, focuses on methods of descriptive statistics. This chapter will focus on using SPSS to produce the descriptive statistics of center, variation, distribution, outliers, and changing characteristics over time. Consult Chapter 3 of *Elementary Statistics*, 10/e or *Essentials of Statistics*, 3/e for a full discussion of the topics.

Section 3-1   Descriptive Statistics

Often descriptive statistics are used to describe characteristics of variables in a data file. The **arithmetic mean** (or **average**), **median,** and **midrange** are measures of the center of a distribution. The **standard deviation**, **variance**, and **range** are measures of the spread of a distribution. **Quartiles** and **percentiles** are measures of position within in a distribution.

Descriptive Statistics for a Variable

Open the Data Set 1: **Health Exam Results** in Appendix B of your textbook (this data is saved on the data disk as **Mhealth**). The **Frequencies** procedure can be used since there are no subgroups in this data file. Choose **Analyze > Descriptive Statistics > Frequencies…** to open the **Frequencies** dialog box (see Figure 2 - 1). Copy the variables *age*, *weight*, and *pulse* to the Variable(s) list and click the **Statistics…** button to open the **Frequencies: Statistics** dialog box (Figure 3 - 1).

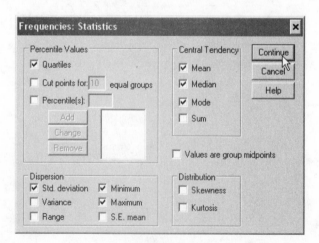

Figure 3 - 1

Check the checkboxes for any descriptive statistics to be included in the analysis. Click the **Continue** button to return to the **Frequencies** dialog box. The frequencies reports will be rather long and so it is probably best to uncheck the checkbox for **Display frequency tables**. Click the **OK** button and the descriptive statistics will appear in the Output Viewer window (Figure 3 - 2).

Statistics

		Ages (years)	Weight (pounds)	Pulse rate (beats/min)
N	Valid	40	40	40
	Missing	0	0	0
Mean		35.48	172.550	69.40
Median		32.50	169.950	66.00
Mode		20	119.5[a]	60[a]
Std. Deviation		13.927	26.3272	11.297
Minimum		17	119.5	56
Maximum		73	237.1	96
Percentiles	25	25.25	152.000	60.00
	50	32.50	169.950	66.00
	75	45.50	190.600	76.00

a. Multiple modes exist. The smallest value is shown

Figure 3 - 2

The descriptive statistics for each variable are placed into one table. This makes it easy to compare the various statistics. SPSS appends a note to the table to indicate that the mode is not unique for waist circumference and pulse rate.

Descriptive Statistics for subgroups of a Variable

Open the **Academy Award Ages** data file. Since there are two subgroups in this data file (indicated by the variable *gender*) the **Explore** procedure will be used. Choose **Analyze > Descriptive Statistics > Explore…** to open the **Explore** dialog box (Figure 2 - 8). Copy *age* into the **Dependent List** box and *gender* into the **Factor List** box. To display descriptive statistics only and suppress the creation of plots choose the bullet for **Statistics**. Click the button for **Statistics…** to open the **Explore: Statistics** dialog box (Figure 3 - 3).

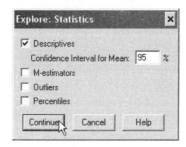

Figure 3 - 3

Choose the checkbox for **Descriptives** and click the **Continue** button. Click the **OK** button and the descriptive statistics will appear in the Output Viewer window. Only a portion of the Descriptives table is shown (Figure 3 - 4) since the output is quite long.

Descriptives(a)

Gender of actor/actress				Statistic	Std. Error
Age of actor/actress	Actress	Mean		35.68	1.269
		95% Confidence Interval for Mean	Lower Bound	33.16	
			Upper Bound	38.21	
		5% Trimmed Mean		34.50	
		Median		33.50	
		Variance		122.406	
		Std. Deviation		11.064	
		Minimum		21	
		Maximum		80	
		Range		59	
		Interquartile Range		12	
		Skewness		1.870	.276
		Kurtosis		4.399	.545
	Actor	Mean		43.95	1.040
		95% Confidence Interval for Mean	Lower Bound	41.88	
			Upper Bound	46.02	
		5% Trimmed Mean		43.52	
		Median		42.00	
		Variance		82.157	
		Std. Deviation		9.064	
		Minimum		29	
		Maximum		76	
		Range		47	
		Interquartile Range		11	
		Skewness		.871	.276
		Kurtosis		1.011	.545

a There are no valid cases for Age of actor/actress when Gender of actor/actress = .000. Statistics cannot be computed for this level.

Figure 3 - 4

The **Explore** procedure calculates several descriptive statistics (mean, median, variance, standard deviation, minimum, maximum, range and interquartile range) for each variable. There is other information in the Descriptives table, which we will put off describing it until we discuss inferential statistics.

Section 3-2   Exercises

1. Refer to Data Set 12: **Weights and Volumes of Cola** in Appendix B (this data is saved on the data disk as **Cola.sav**). The data set contains the weight and volume measurements on 36 cans of Regular Cola, Diet Cola, Regular Pepsi, and Diet Pepsi.

   a. Determine the mean and standard deviation of the weights of Regular Cola, Diet Cola, Regular Pepsi, and Diet Pepsi. What can you conclude about the weights of Colas?

   b. Find the 5-number summary for the weights of Regular Cola, Diet Cola, Regular Pepsi, and Diet Pepsi.

2. *Age of Presidents.* A Senator is considering running for the U.S. presidency, but she is only 35 years of age, which is the minimum required age. While investigating this issue, she finds the ages of the past presidents when they were inaugurated, and those ages are listed in Table 3 - 1.

57	61	57	57	58	57	61	54	68	51	49
64	50	48	65	52	56	46	54	49	51	47
55	55	54	42	51	56	55	51	54	51	60
52	43	55	56	61	52	69	52	64	46	54

Table 3 - 1

Using the listed ages, find the
a. Mean,
b. Median,
c. Mode,
d. Range,
e. Standard deviation,
f. Variance,
g. Q_1 and Q_3,
h. and P_{10}.

3. *Old Faithful Geyser.* The table below lists the intervals (in minutes) between eruptions of the Old Faithful Geyser in Yellowstone National Park.

98	92	95	87	96	90	65	92	95	93	98	94

Using the listed intervals, find the
a. Mean,
b. Median,

        c. Mode,

        d. Range,

        e. Standard Deviation,

        f. and Variance

4. *Ages of Stowaways.* The *Queen Mary* sailed between England and the United States, and sometimes stowaways were found on board. The following table lists the ages for two data sets, one for the eastbound crossing and one for the westbound crossing. Compare the two data sets.

Eastbound	24	24	34	15	19	22	18	20	20	17
Westbound	41	24	32	26	39	45	24	21	22	21

Chapter 4

Probability

Chapter 4   Probability

Chapter 4 of *Elementary Statistics*, 10/e and *Essentials of Statistics*, 3/e primarily focus on the theoretical calculations of probabilities. Section 4-6 of discusses calculating probabilities through simulation. You should be familiar with this section prior to beginning this chapter.

For many problems, it is impossible, very difficult, or impractical to determine a particular probability from theoretical calculations. In situations like this, simulation is often used to estimate probabilities. **Simulation** uses randomness to artificially duplicate an experiment to determine its effect.

For example, consider estimating the probability that among the next twenty births at a particular hospital there will be more girls born than boys born. We know that about half of births are girls and about half are boys. A single birth can be simulated by a coin flip; a girl birth could be recorded if the coin lands heads and a boy birth if the coin lands tails. To simulate twenty births, twenty coins could be flipped. The number of heads and tails simulate the number of girl and boy births. Finally, the probability of more girls than boys being born can be estimated by repeating this simulation a large number (e.g. *1000*) of times. The proportion of the *1000* simulations that result in more heads (girls) than tails (boys) is an estimate of the probability.

Consider estimating the probability of more girls than boys in the next ten thousand births. Using coins to estimate this probability would be very laborious and time consuming. For this reason, computers are used to compute random numbers for simulation problems.

This chapter begins with a discussion of random numbers and their properties. Section 4-2 describes how to calculate random numbers in SPSS. Then a few examples of simulations are given.

Section 4-1   Random Digits and Random Numbers

Random digits are integers produced in such a way that each of the digits *0* through *9* has probability one-tenth of occurring. Each pair of digits *00* through *99* has probability one-hundredth of occurring. Further, each triple of digits *000* through *999* has probability one-thousandth of occurring and so on. Random digits tables have the property that every number in the table has the same probability of occurring.

There are tables of Random Digits. The book, *A Million Random Digits with 100,000 Normal Deviates* published by Rand Corporation, Glencoe, Illinois, The Free Press in *1955* is one of the largest sources of random digits. Tables such as this one are rarely used anymore because computers can calculate thousands of random digits in microseconds.

Simulation problems often require random numbers that have special properties. We have already seen that random digits have the property that every group of *k*-digits has the same probability of occurring. Random digits are discrete; many times continuous random numbers are needed. Random numbers that take on values between *0* and *1*, and are created in such a way that every sub-interval of length *p* has probability *p* of occurring are commonly used in simulations. Random numbers with this property are called Uniform random numbers on the interval *(0, 1)*.

There are many different properties that random numbers can have. For example, random numbers that have a particular mean and standard deviation might be required. In the next chapter, we will learn about probability distributions. Often, random numbers that have a particular probability distribution are required. SPSS can compute random numbers with over twenty different kinds of distributions.

Section 4-2   Calculating Uniform Random Numbers

Uniform random numbers on the interval *(0, 1)* are quite useful in simulations. For the above example, a random number between zero and one-half could represent a girl birth and a random number between one-half and one could represent a boy birth. More generally, a random number between zero and *p* could represent an event that happens with probability *p*.

To calculate *20* Uniform random numbers on the interval *(0, 1)* in SPSS, create a new data file by choosing **File > New > Data**. Before we begin, we need to create a new variable with *20* cases. It is unimportant what the values of the cases are; SPSS will create a random variable for each case in the data file. Create a variable named *uniform* (choose the **Variable View** tab, if the Data Editor is in Data View) and then type *uniform* into the first row of the column labeled **Name** (for more details see Section 0-4 of this manual). Click the **Data View** tab and scroll down to the *20*th case and enter any value (say *1*). The variable *uniform* will now have *20* cases in it; the first *19* cases will have a dot or period (denoting a missing value) in them and the *20*th case will have the value *1*.

To calculate the random numbers, choose **Transform > Compute…** to open the **Compute Variable** dialog box (Figure 4 - 1). We will use the **RV.UNIFORM(min, max)** function to create the uniform random numbers. Many different functions begin with "RV." in the Functions list. All the functions that begin with "RV." generate random numbers. The different functions generate random numbers with different properties. **RV.UNIFORM(min, max)** returns a uniformly distributed random number greater than the argument *min* and less than the argument *max*. In order to generate numbers on the interval *(0, 1)*, we use **RV.UNIFORM(0, 1).** Type *uniform* into the **Target Variable** box and click on **Random Numbers** in the **Function Group** box, then scroll down in the **Functions and Special Variables** box and click on **Rv.Uniform** and then click the **Paste Expression** ▲ button. This will paste *RV.UNIFORM(?, ?)* into the **Numeric Expression.** Note that when we click on the function, information about the function appears in a box. We require random numbers between 0 and 1, therefore edit the expression to read *RV.UNIFORM (0, 1)*. The **Compute Variable** dialog box should look like Figure 4 - 1.

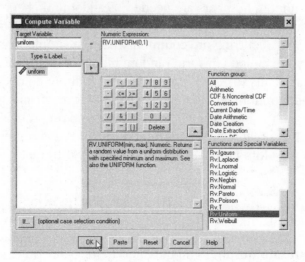

Figure 4 - 1

Click the **OK** button and SPSS will ask if it is OK to "**Change the existing variable?**". This is because we created the variable *uniform* with *20* cases and now we want to store the random numbers in the same variable. SPSS will always ask if it is OK to change a variable before overwriting it. Click the **OK** button and SPSS will compute *20* **Uniform random numbers** on the interval *(0, 1)* and store them in the variable named *uniform*. You might want to change the format of *uniform* from *Numeric8.2* to display more digits. The process that created the random numbers guarantees that the random numbers have the property, that every sub-interval of the interval *(0, 1)* has the same probability of occurring. Simulation exploits this property to estimate probabilities.

Section 4-3   Simulating Probabilities

Consider the problem posed in the introduction to this chapter of simulating the probability that among the next twenty births at a particular hospital there will be more girls than boys. Because of the way the variable *uniform* was created, it follows that half the time the value of *uniform* will be between zero and one-half and half of the time the value of uniform will be between one-half and one. We can simulate twenty births by denoting girl births by values of *uniform* that are between zero and one-half and denoting boy births by values of *uniform* that are between one-half and one. Counting the number of values less than one-half gives the number of girl births.

Counting the number of data values less than one-half can be made easier by sorting the data values of the variable *uniform* from smallest to largest. Choose **Data > Sort Cases...** to open the **Sort Cases** dialog box (Figure 4 - 2).

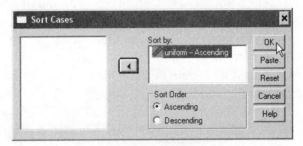

Figure 4 - 2

Paste *uniform* in the **Sort by** box and choose the bullet for **Sort Order Ascending** (smallest to largest) then click the **OK** button. The variable *uniform* is now sorted from smallest to largest in the Data Editor. Scroll down the Data Editor to see how many of the values are less than *0.50*, equivalently, this is the number of girls that the simulation estimates will be born in the next twenty births. This is the same as flipping twenty coins once. To estimate the probability that more girls are born than boys, we will need to repeat the process a large number of times.

We can quickly simulate another *20* births by choosing **Transform > Compute...** again. The **Compute Variable** dialog box should still have the information in it from when the variable *uniform* was created; it remembers what we had typed in the previous time. This time type *uniform2* into the **Target Variable** box and then click the **OK** button. The Data Editor now has two variables, *uniform* and *uniform2*.

Create eight more variables *uniform3*, *uniform4*, ... *uniform9*, *uniform0* in the same way. Sort each variable from smallest to largest and determine if there are more girls born than boys born (that is, eleven or more girls). Once the variable is sorted if the *11*[th] case is less than *0.50* then more girls than boys were born. Notice that each time you sort a variable the cases of the other variables change order as well. This is because in SPSS the cases for each variable are related. The proportion of the ten variables with more girls than boys is an estimate of the probability that we seek. Your answer should be near *0.41* but after only simulating this process *only* ten times there will be a good deal of variation in your answers. It is very likely that you are thinking that SPSS is not really simplifying the problem since this process is complicated. The idea of repeating this process *10,000* times to obtain a less variable result seems to be out of the question.

In the next chapter, we will learn that the number of girls born out of a sample of *20* births will follow a certain **Binomial probability distribution**. SPSS can generate random numbers from a **Binomial probability distribution**. In this case, each random number will represent the number of girls born in twenty births. Using random numbers from a Binomial probability distribution will allow us to quickly simulate this experiment a large number of times.

Open a new data file (choose **File > New > Data**); the data file above will not be used again in this manual so there is no need to save it. Create a variable named *binomial* with *5000* cases. The procedure we used in Section 4-2 was sufficient for a small data file containing 20 cases. In SPSS 14.0, only enough memory is allocated for an empty file of the approximate size to fill one screen, or about 40 cases. SPSS provides a powerful command language that allows you to save and automate many common tasks. A command syntax file is simply a text file containing SPSS commands to accomplish certain tasks. Open the syntax editor (choose **File > New > Syntax**) and type in the code shown in Figure 4 – 3.

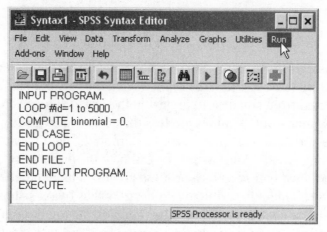

Figure 4 - 3

To "run" this command file, simply click **Run > All**. When these statements execute, *5000* cases will be generated with each case of *binomial* set to zero. It does not matter what value we set each case to; we are simply assigning an arbitrary value of zero to be able to identify the cases more easily. To convince yourself that all the cases have been allocated, you can scroll through the cases with the Page Down key until you see the 5000th case of *binomial*. You can also open the **Go to Case** dialog box by choosing **Edit > Go to Case…** Type in *5000* and click the **OK** button and the Data Editor will select the 5000th case of binomial. Since we are interested in generating integer values, change the format of *binomial* to *Numeric8.0*. You can find out more about the SPSS command language by consulting **Help** or referring to your reference manuals which are stored as PDF files and provided by the vendor.

Choose **Transform > Compute…** to open the **Compute Variable** dialog box. Type *binomial* into the **Target** Variable box and click on **Random Numbers** in the **Function Group** box, then scroll down in the **Functions and Special Variables** box and click on **RV.BINOM(n, prob)** and then click the **Paste Expression** ▲ button. The function *RV.BINOM(n, prob)* calculates random numbers that come from a Binomial distribution with parameters *n* and *prob* (you will learn more about this distribution in the next chapter).

Edit the **Numeric Expression** so that it reads *RV.BINOM(20, .5)*. The first parameter *20* is the number of births and the second parameter is the probability of a girl baby. The **Compute Variable** dialog box should look like Figure 4 - 4.

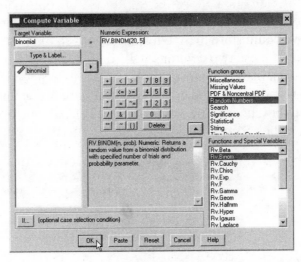

Figure 4 - 4

Click the **OK** button and SPSS will ask if it is OK to "**Change the existing variable?**". Again, this is because we created the variable *binomial* with *5000* cases and now we want to store the random numbers in the same variable. SPSS will always ask if it is OK to change a variable before overwriting it. Click the **OK** button and the *5000* random numbers will show up in the Data Editor. Sort the variable *binomial* from smallest to largest (ascending). Then scroll down the Data Editor until you see a data value of eleven; in my Data Editor, the first eleven occurs at the *2942*nd case. Therefore, there were *2941* simulations (cases) with *10* or fewer girls and *5000* – *2941* = *2059* simulations with *11* or more girls. It follows for my data, that *2059* of the *5000* (or *41.1%*) simulations had more girls than boys. The simulation estimates the probability that more girls than boys are born in twenty births to be *0.411*. This is very close to the true value of *41%*.

Section 4-4   Simulating the Sum of Two Dice

Consider the example in Section 4 - 6 of your textbook that describes a procedure for simulating the rolling of a pair of dice. When rolling two dice, each of the numbers *1* through *6* happens with the same chance on each die. If the two numbers that result from the roll of the two dice are added together then the resulting value values will be between 2 and 12, inclusively. Below we will discuss how to simulate the sum of two dice.

We can simulate the process of rolling a single die by generating **uniform random numbers** on the interval (*1, 7*). These random numbers are continuous and the six intervals (*1, 2*), (*2, 3*), …, (*6, 7*) are all equally likely because the intervals are the same length. If we truncate each of the random numbers to an integer the result will be to create random integers (digits between 1 and 6, inclusive) that are equally likely.

To simulate the sum of the two dice, we simulate the process of rolling a single die twice. We could store these two simulations in the two variables, *die1* and *die2*. Then the sum of these two columns will simulate the sum of the two die. Let's simulate the sum of *1000* pairs of dice.

Open a new data set, and create two new variables named *die1* and *die2*. Open the syntax editor (choose **File > New > Syntax**) and type in the code shown in Figure 4 - 5 and then click **Run > All**.

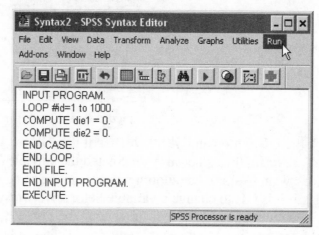

Figure 4 - 5

This will create *1000* cases with each case of *die1* set to zero and *1000* cases with each case of *die2* set to zero. We are simply allocating storage for a data file having *1000* cases and setting each case of the variable *die1* and *die2* to an arbitrary value of zero. Set the format for *die1* and *die2* to *Numeric8.0*.

To simulate rolling the first die, choose **Transform > Compute…** to open the **Compute Variable** dialog box (Figure 4 - 4). In the **Target Variable** box, type the variable name *die1*. In the **Numeric Expression** box, type the expression *TRUNC(RV.UNIFORM(1, 7))*. This expression uses two SPSS functions **TRUNC(numexpr)** in the arithmetic function group and **RV.UNIFORM(min, max)** in the random numbers function group. When you click on either of these functions, a description of that the function appears in a window. Click the **OK** button to generate the random numbers and **OK** to change the existing variable. Repeat the compute process for *die2* to simulate the roll of the second die.

Now create a new variable *sum* by adding the two variables together. To do this, choose **Transform > Compute…** to open the **Compute Variable** dialog box. Type *sum* into the **Target Variable** box and then type, *die1 + die2*, into the **Numeric Expression** box. Click the **OK** button to calculate the variable *sum* and **OK** to change the existing variable. You can scroll through the Data Editor to see that the variable *sum* is the sum of the two variables, *die1* and *die2*.

Make a frequencies report and bar chart for the variable *sum*. Choose **Analyze > Descriptive Statistics > Frequencies…** to open the **Frequencies** dialog box (Figure 2 - 1). Select the variable *sum* into the **Variable(s)** box. To add a bar chart to the frequencies report, click the **Charts…** button to open the **Frequencies: Charts** dialog box (Figure 4 - 6). Choose the bullet for **Bar charts** and click the **Continue** button.

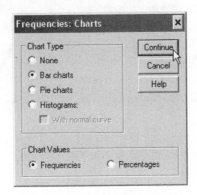

Figure 4 - 6

Click the **OK** button and the frequencies report (Figure 4 - 7) for sum will appear in the Output Viewer window. We can see that a sum of *10* happened *85* times in the *1000* simulations. That means that the sum was exactly ten *8.5%* of the time. We can also see that the sum is seven or less about *59.4%* of the time.

sum

		Frequency	Percent	Valid Percent	Cumulative Percent
Valid	2	30	3.0	3.0	3.0
	3	68	6.8	6.8	9.8
	4	77	7.7	7.7	17.5
	5	105	10.5	10.5	28.0
	6	137	13.7	13.7	41.7
	7	177	17.7	17.7	59.4
	8	114	11.4	11.4	70.8
	9	114	11.4	11.4	82.2
	10	85	8.5	8.5	90.7
	11	56	5.6	5.6	96.3
	12	37	3.7	3.7	100.0
	Total	1000	100.0	100.0	

Figure 4 - 7

The bar chart (Figure 4 - 8) for *sum* appears below the frequencies report in the Output Viewer window. In the next chapter, we will learn that bar charts tell us about the distribution of data values. We can see here that the most likely value for the distribution of the sum is 7 and that the distribution is roughly symmetric.

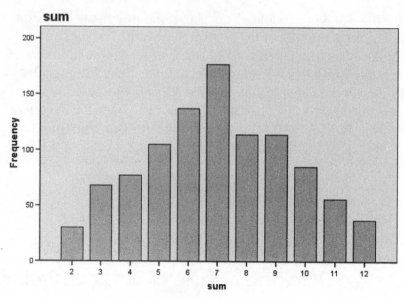

Figure 4 - 8

Section 4-5   Exercises

1. Calculate a random sample of *100* **Uniform random numbers** on the interval (*0, 10*). Make a histogram of the distribution of these *100* random numbers. Do not make a frequencies report for this data since it would be very long.

2. Calculate a random sample of *100* **Uniform random digits** between *1* and *10*, inclusive. Make a frequencies report and a bar chart of the distribution of these *100* random digits.

3. Suppose that the *60%* of births are girls and *40%* are boys. Generate a random sample of *1000* random numbers from a **Binomial probability distribution** with *n = 20* and *p = 0.6*. Simulate *1000* births, to estimate the probability that more girls than boys are born in twenty births.

4. Simulate rolling three dice *100* times. Use your simulation to estimate the probability of observing a sum of *10* when three dice are rolled.

Chapter 5

Probability Distributions

Chapter 5   Probability Distributions

Chapter 4 of *Elementary Statistics*, 10/e and *Essentials of Statistics*, 3/e introduce discrete random variables and their distributions. A **random variable** is a variable (typically represented by x) that has a single numerical value, determined by chance, for each outcome of a procedure. Such as the sum of two dice that result when a pair of dice is rolled, the number of cars passing through an intersection each hour, or the lifetime of a flashlight battery.

A **discrete random variable** has either a finite number of values or a countable number of values. Countable means that there may be infinitely many values but they can be listed even though the list may continue indefinitely. SPSS can calculate probabilities for over twenty-five distributions. This chapter discusses calculating probabilities for discrete random variables that have either a **Binomial probability distribution** or a **Poisson probability distribution**. Both of these distributions occur frequently in statistics.

Section 5-1   Binomial Probability Distributions

The **binomial probability distributions** are important because they allow us to deal with circumstances in which the outcomes belong to one of two mutually exclusive categories, such as defective and acceptable or yes and no responses to a question. Notice that the number of girls in *20* births (the probability we simulated in Section 4-3) involves two mutually exclusive categories, boys and girls, and thus falls into this situation. Be aware that not all two-category problems are solved by binomial probability distributions.

A **binomial probability distribution** results when a procedure satisfies *all* four of the following requirements:

1. The procedure has a *fixed number of trials*. The number of trials is represented by n.
2. The trials are *independent*. That is the outcome of any individual trial doesn't affect the probabilities in the other trials.
3. Each trial must have all outcomes classified into one of *two categories*. The two categories are sometimes referred to as success (S) and failure (F).
4. The probabilities of the two categories remain *constant* for each trial. The probability of a success is represented by p and the probability of a failure is then $q = 1 - p$.

A random variable x is said to have a **binomial probability distribution** with parameters n and p, if x is the number of successes in the n trials. Chapter 5 of your textbook gives the binomial probability formula for calculating the probability that there are x successes in the n trials as

$$P(x) = \frac{n!}{(n-x)!\,x!} \cdot p^x \cdot q^{n-x} \qquad \text{for } x = 0, 1, 2, \ldots, n$$

where n is the number of trials, x is the number of successes among the n trials, and p is the probability of success on any one trial.

The above formula gives the probability for finding the probability of exactly x successes in n trials, but sometimes it is required to find the probability of x or fewer successes. The probability of x or fewer successes is called the **cumulative probability distribution**. SPSS calculates cumulative probabilities. The probability of exactly x successes can be obtained from the cumulative probability function.

Consider finding the probability of more girl births than boy births in 20 births. We estimated this probability to be 0.411 in Section 4-3 of this manual. Now we will calculate the exact probability. Notice that all four requirements for a binomial distribution are satisfied– there are a fixed number of trials (the 20 births), the births are independent, only girls or boys are possible, and the chance of having a girl is one-half for each birth. Therefore, the number of girl births in the next 20 births has a binomial probability distribution with $n = 20$ (number of trials) and $p = 0.50$ (probability of a girl birth). We will have SPSS calculate some probabilities for this problem.

First, we calculate the probability that there are less than or equal 7 girl births in the next 20 births. Choose **Transform > Compute…** to open the **Compute Variable** dialog box (Figure 5 - 1). Note an error may occur if there is no data in the Data Editor. This problem does not require any data; therefore simply enter any number into any cell in the Data Editor. The function **CDF.BINOM(x, n, p)** computes the probability of x or fewer successes for a binomial probability distribution with parameters n and p. Type the expression *CDF.BINOM(7, 20, 0.5)* into the **Numeric Expression** box and type *prob* into the **Target Variable** box (see Figure 5 - 1).

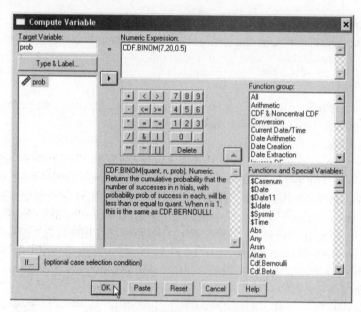

Figure 5 - 1

Click the **OK** button and SPSS will calculate the probability of *7* or fewer girl births in the next *20* births to be *0.131588* (the Data Editor will show the value of *.13*, since the default data type in SPSS is *Numeric8.2*, which displays only two decimal places).

SPSS can calculate the probability of exactly *x* successes as well. For example, we can calculate the probability of exactly *7* girl births in the next *20* births. The probability of exactly *7* births is found by subtracting the probability of *6* or fewer girl births from the probability of *7* or fewer girl births. In SPSS we calculate this probability by choosing **Transform > Compute…** to open the **Compute Variable** dialog box and typing the expression

CDF.BINOM(7, 20, 0.5) – CDF.BINOM(6, 20, 0.5),

into the **Numeric Expression** box. Change the **Target Variable** to *prob2* and click the **OK** button. The required probability is *0.073929*.

Recall our interest in this problem was to calculate the probability that there are more girl births than boy births in the next twenty births. This is equivalent to the probability that the number of girl births is at least *11* (that is, *11* or more). Since all the probabilities for a probability distribution add to one, we can find the probability of at least *11* girl births by subtracting the probability of ten or fewer girl births from *1*.

Determine the probability that at least *11* girl births. Choose **Transform > Compute…** and edit the **Numeric Expression** box to read *1 - CDF.BINOM(10, 20, 0.5)*. Change the **Target Variable** to *prob3* and click the **OK** button. The required probability is *0.411901*.

Notice that this is the exact probability that there will be more girl births than boy births in the next twenty births. In Section 4-3 of this manual, we simulated the probability of more girl births than boy births and found *0.411*. The probability we found in Section 4-3 is only an estimate of the exact probability. If we were to simulate the probability in Section 4-3 again, we would get a different value, though one that is very close to the true value of *0.41* that we found in this section.

Section 5-2   Obtaining a Table of Binomial Probabilities

Sometimes the table of probabilities for a binomial probability distribution is required. Table A-1 in your textbook has some tables of binomial probabilities. Table A-1 has values of *n* from *2* to *15* for thirteen values of *p* (namely, *.01, .05, .10, .20, .30, .40, .50, .60, .70, .80, .90, .95,* and *.99*). This section shows you how to calculate a table of binomial probabilities for any value of *n* and *p*.

Problems 25-28 in Section 5-3 of your textbook, require a table of binomial probabilities with $n = 6$ and $p = 0.167$. To determine the table of probabilities for the binomial probability distribution with parameters $n = 6$ and $p = 0.167$, first create a variable named x that has the numbers 0 through *n*, in this situation *6*, in the first 7 cases of the variable x.

Since SPSS computes cumulative probabilities, it is a little more work to obtain the table of *exact* probabilities. Choose **Transform > Compute...** to open the **Compute Variable** dialog box (Figure 5 - 1). The procedure is similar to the way we calculated the probability of exactly *7* girl births in the example above. Type
$$CDF.BINOM(x, 6, 0.167) - CDF.BINOM(x-1, 6, 0.167)$$
into the **Numeric Expression** box and type *prob* into the **Target Variable** box. Click the **OK** button to obtain the probability table for this binomial distribution. Change the data type of the variable ***prob*** from *Numeric8.2* to *Numeric8.4* and compare the Data Editor (Figure 5 - 2) with the table of probabilities shown in your textbook.

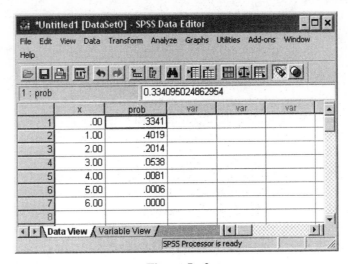

Figure 5 - 2

Section 5-3   Poisson Probability Distributions

The **Poisson probability distributions** are particularly important distributions because they are often used as a mathematical model for describing behavior such as radioactive decay, planes arriving at an airport, and Internet users logging onto a website. The **Poisson probability distribution** is a discrete probability distribution that gives the probability of x occurrences of some event *over a specified interval*. The probability of the event occurring x times over an interval is given by the formula

$$P(x) = \frac{\mu^x \cdot e^{-\mu}}{x!} \qquad \text{for } x = 0, 1, 2, \ldots$$

where e is Euler's constant, which is approximately *2.718282*. The parameter μ gives the mean number of occurrences of the event in the specified interval. SPSS can be used to calculate cumulative probabilities in the same way that binomial probabilities were calculated.

For example, suppose your professor has an office hour scheduled every Monday at 11:00 and she finds that during that office hour, the mean number of students who visit is *2.3*. We can use the Poisson probability distribution to determine probabilities concerning the number of students that visit a randomly selected office hour on Monday at 11:00.

To compute the probability of *4* or fewer students visiting during a randomly selected office hour, choose **Transform > Compute…** to open the **Compute Variable** dialog box (Figure 5 - 1). The function **CDF.POISSON(x, mean)** computes the probability of x or fewer successes for a Poisson probability distribution with the given mean. Type the expression *CDF.POISSON(4, 2.3)* into the **Numeric Expression** box and type *prob* into the **Target Variable** box. Click the **OK** button and the probability of *4* or fewer occurrences is found to be *0.916249*.

To find the probability that *exactly 4* students will visit during a randomly selected office hour, type the expression *CDF.POISSON(4, 2.3) – CDF.POISSON(3, 2.3)* into the **Numeric Expression** box. Click the **OK** button and the probability that exactly four students will come to her office is found to be *0.116902*.

Section 5-4   Approximating Binomial Probabilities by a Poisson

The Poisson probability distribution is sometimes used to approximate binomial probabilities when n is large and p is small, for example when $n \geq 100$ and $\mu = np \leq 10$. When these conditions are satisfied the approximate Poisson probability is very close to the exact binomial probability.

We will use the following example to verify that the approximate Poisson probability is very close to the binomial probability. In Illinois' Pick 3 game, you pay *50* cents to select a sequence of three digits, such as 911. If you play this game once every day for a year, find the probability of winning exactly one time during the year (*365* days).

Because the time interval is *365* days, *n = 365*. Because there is one winning set of numbers among the *1000* (000 to 999) that are possible, *p = 1/1000 = 0.001*. The conditions for using the Poisson distribution to approximate binomial probabilities are satisfied (namely, $n = 365 \geq 100$ and $\mu = np = 365 \times 0.001 = 0.365 \leq 10$).

The Poisson probability of exactly one win during the year is found by typing
$$CDF.POISSON\ (1,\ 0.365) - CDF.POISSON(0,\ 0.365)$$
into the **Numeric Expression** box of the **Compute Variable** dialog box. The probability is found to be *0.253381*.

The exact binomial probability of exactly one win during the year is found by typing
$$CDF.BINOM(1,\ 365,\ 0.001) - CDF.BINOM(0,\ 365,\ 0.001)$$
into the **Numeric Expression** box of the **Compute Variable** dialog box. The probability is found to be *0.253589*. The Poisson probability differs from the binomial probability by only 0.000207. The approximate Poisson probability is very close to the exact binomial probability.

Section 5-5   Exercises

1. A student guesses the answer to all the questions on a *20* question multiple-choice exam. The exam had four possible answers for each question. A passing grade on the exam is *60%* or *12* questions. What is the probability that the student passes the exam?

2. The Teletronic Company purchases large shipments of fluorescent bulbs and uses the following acceptance-sampling plan. Randomly select and test *24* bulbs, then accept the whole batch if there is only one or none that do not work. If a particular shipment of thousands of bulbs actually has a *4%* rate of defects then what is the probability that this whole shipment will be accepted?

3. Consider an experiment in which a die is rolled *100* times

    a. find the probability of *1* occurring *20* or less times in *100* rolls of a *fair* die.

    b. find the probability of *1* occurring exactly *20* times in *100* rolls of a *fair* die.

4. Currently, *11* babies are born in the village of Westport (population 760) each year (based on data from the U.S. National Center for Health Statistics). Use SPSS to

    a. find the probability that on a given day, there are no births.
    b. find the probability that on a given day, there is at least one birth.

5. A virus is known to infect a human only *0.1%* of the time and the virus is found in the air conditioning system of a large hotel. If there are *10,000* people who are exposed to the virus at the hotel, what is the probability that more than *20* people become infected?

6. For a recent period of *100* years, there were *93* major earthquakes (at least *6.0* on the Richter scale) in the world (based on data from the *World Almanac and Book of Facts*). If the Poisson distribution is a suitable model, use SPSS to find the probability that there is *1* earthquake in a selected year.

Chapter 6

Normal Probability Distributions

Chapter 6   Normal Probability Distributions

Normal probability distributions are the most important and widely used distributions in statistics. It is often appropriate to use a normal probability distribution to describe a random variable. Furthermore, the normal distribution is applied extensively in inferential statistics. Most of the remaining topics in *Elementary Statistics*, 10/e and *Essentials of Statistics*, 3/e are based on the normal probability distribution. In this chapter, we will learn how to calculate probabilities from a normal distribution and how to calculate the score that corresponds to a probability. Further, we will learn how to generate sample data from a normal probability distribution and to determine if a data set follows a normal probability distribution. Section 6-5 of this manual on determining normality is not covered in *Essentials of Statistics*. You should be familiar with Chapter 6 of *Elementary Statistics*, 10/e or *Essentials of Statistics*, 3/e prior to beginning this chapter.

Section 6-1   Finding Probabilities Given Scores

A **normal probability distribution** is a bell-shaped symmetric curve with mean, μ, and standard deviation, σ. The normal distribution is symmetric about the mean. The standard deviation of the normal distribution determines how spread out the distribution is around the mean. The total area under the normal curve is equal to *1*. A special case of the normal distribution is the **Standard Normal** probability distribution, which has mean *0* and standard deviation *1*.

Following the notation in Chapter 6 of your textbook, *P(a < z < b)* denotes the probability that the *z* score is between *a* and *b*; *P(z < a)* denotes the probability that the *z* score is less than *a*; and *P(z > a)* denotes the probability that the *z* score is greater than *a*. When calculating probabilities by hand it is often necessary to convert the values in a probability statement to *z* scores (scores associated with a standard normal distribution) and then look up the associated probability in the standard normal table (Table A-2 in Appendix A of your textbook). When calculating probabilities in SPSS it is unnecessary to convert the values to *z* scores.

The function **CDFNORM(a)** computes the probability that a standard normal random variable (mean *0* and standard deviation *1*) is less than *a*, that is *P(z < a)*. The more general function **CDF.NORMAL(a, mean, stdev)** computes the probability that a normal random variable with given mean and standard deviation is less than *a*. The function CDF.NORMAL(a, 0, 1) is equivalent to CDFNORM(a). This manual will use the more general function in the examples that follow.

To calculate *P(z < 1.23)*, create a new variable named *a*, that has the value *1.23* entered into the first cell. Choose **Transform > Compute...** to open the **Compute Variable** (see Figure 4 - 1) dialog box. Type the expression *CDF.NORMAL(a, 0, 1)* into the **Numeric Expression** box and type *prob* into the *Target Variable* box. Click the **OK** button and SPSS will calculate the probability to be *0.890651*. You may have to change the format of the variable *prob* to see all the digits. Compare this value with the result found using Table A-2 in Appendix A of your textbook.

The function CDF.NORMAL always calculates the probability that the score is less than the quantity *a;* therefore, every problem must be written in terms of probabilities that are less than a given quantity. For example, to calculate the probability that a *z* score is more than *a*, you must use the fact that $P(z > a) = 1 - P(z < a)$. Calculate the probability $P(z > 2.34)$, by choosing **Transform > Compute...** to open the **Compute Variable** dialog box. Type *prob2* into the **Target Variable** box and the expression $1 - CDF.NORMAL(2.34, 0, 1)$ into the **Numeric Expression** box. Click the **OK** button and SPSS will report the probability to be *0.009642*. Compare this value with the result found using Table A-2 in Appendix A of your textbook.

Likewise, to calculate the probability $P(-0.12 < z < 2.34)$, first write the probability in the form $P(z < 2.34) - P(z < -0.12)$. That is, the probability that the *z* score is between *-0.12* and *2.34*, is calculated by finding the probability that the *z*-score is less than *2.34* and subtracting the probability that the *z*-score is less than *-0.12*. To calculate this probability, choose **Transform > Compute...** to open the **Compute Variable** dialog box. Type the expression

$$CDF.NORMAL(2.34, 0, 1) - CDF.NORMAL(-0.12, 0, 1)$$

into the **Numeric Expression** box. Type *p* into the **Target Variable** box and click the **OK** button. SPSS will report the probability to be *0.538117*.

Consider the *IQ Scores* Example in Chapter 6-3 of your textbook. A psychologist is designing an experiment to test the effectiveness of a new training program for airport security screeners. The test will be conducted using a homogeneous group of subjects having IQ scores between *85* and *125*. Given that the IQ scores are normally distributed with a mean of *100* and a standard deviation of *15*, what percentage of people have IQ scores between *85* and *125*, or equivalently what is the probability $P(125 < x < 85)$, or equivalently what is $P(x < 125) - P(x < 85)$? To calculate this probability, choose **Transform > Compute...** to open the **Compute Variable** dialog box. Type the expression

$$CDF.NORMAL(125, 100, 15) - CDF.NORMAL(85, 100, 15)$$

into the **Numeric Expression** box and name the **Target Variable** *p*. Click the **OK** button and SPSS will calculate the probability as *0.793554*. Compare with the result in your textbook. The difference between the two results is that when calculating the *z* scores by hand the values were rounded to two decimal places. SPSS did no rounding in the calculations.

In order to have a precision dance team with a uniform appearance, height restrictions are placed on the famous Rockette dancers at New York's Radio City Music Hall. Because women have grown taller, a recent change requires that a Rockette dancer must have a height between *66.5* inches and *71.5* inches. If a woman is randomly selected, find the probability that she will meet the height requirement to be a Rockette. In other words, the probability $P(66.5 < x < 71.5)$ is required. It is assumed that the heights of women are normally distributed with a mean of *63.6* inches and a standard deviation of *2.5* inches (based on data from a National Health Survey). To calculate this probability, choose **Transform > Compute...** to open the **Compute Variable** dialog box. Type the expression

   CDF.NORMAL(71.5, 63.6, 2.5) – CDF.NORMAL(66.5, 63.6, 2.5)

into the **Numeric Expression** box and name the **Target Variable** *p*. Click the **OK** button and SPSS will calculate the probability as *0.122236*. Only about *12%* of women meet the Rockette's new height requirement.

Section 6-2   Finding Scores Given Probabilities

In the previous section, the examples involved finding a probability corresponding to a given score. This section is concerned with the reverse or inverse of the problem of the previous section. We now want to find the score that corresponds to a given probability. It is easy to confuse the probability and the score in these problems. Drawing a picture will help keep everything straight. Remember that scores are distances along the horizontal axes and the probabilities are areas (whose values are always between *0* and *1*) associated with regions under a normal curve.

   We saw in the previous section, that the function CDF.NORMAL computed probabilities given scores. Now we learn that the function IDF.NORMAL computes scores given probabilities. **IDF.NORMAL(p, mean, stdev)** calculates the score from a normal distribution, with specified mean and standard deviation, for which the cumulative probability is *p*.

   Let's see how to determine the 95th percentile of a standard normal distribution. That is, find the *z* score that has probability *0.95* to the left of it and probability *0.05* to the right of it; or equivalently, solve the equation $P(z < a) = 0.95$ for *a*.

   To find this *z* score, choose **Transform > Compute...** to open the **Compute Variable** dialog box. Type the expression *IDF.NORMAL(0.95, 0, 1)* into the **Numeric Expression** box and name the **Target Variable** *p*. Click the **OK** button and SPSS will calculate the *z* score as *1.644854*. This is the exact value of the *z* score; your textbook rounds this value to *1.645*.

   According to a National Health Survey, the heights of women are normally distributed with a mean of *63.6* inches and a standard deviation of *2.5* inches. Find the 90th percentile of women's heights, that is find the height that separates the top *10%* of heights from the bottom *90%* of heights. The problem is to find the score that solves the equation $P(x < a) = 0.90$. To find this score, choose **Transform > Compute...** to open the **Compute Variable** dialog box. Type the expression *IDF.NORMAL(0.90, 63.6, 2.5)* into the **Numeric Expression** box and name the **Target Variable** *a*. Click the **OK** button and SPSS will calculate the score as *66.80388* inches. Notice that we did not have to compute a *z* score to find this value.

Section 6-3   Approximating Binomial Probabilities by a Normal

A normal probability distribution is sometimes used to approximate binomial probabilities when n is large, for example when both $np \geq 5$ and $n(1 - p) \geq 5$. When both these conditions are satisfied a normal probability distribution with mean $\mu = np$ and the standard deviation is $\sigma = \sqrt{np(1-p)}$ will give probabilities that are very close to the exact binomial probabilities.

We will use the following example to verify that a normal probability distribution can be used to approximate binomial probabilities. When an airliner is loaded with passengers, baggage, cargo, and fuel, the pilot must verify that the gross weight is below the maximum allowable limit, and the weight must be properly distributed so that the balance of the aircraft is within safe acceptable limits. Air America has established a procedure whereby extra cargo must be reduced whenever a plane filled with *200* passengers includes at least *120* men. Find the probability that among *200* randomly selected passengers, there are at least *120* men. Assume that the population of potential passengers consists of equal numbers of men and women.

The number of male passengers follows a binomial distribution with a fixed number of trials ($n = 200$), which are presumably independent, with two categories (men and women) of outcome for each trial, and a probability of a male ($p = 0.5$) that presumably remains constant from trial to trial. The conditions for using the normal distribution to approximate binomial probabilities are satisfied (namely, $np = 200 \times 0.5 = 100$ and $n(1 - p) = 200 \times 0.5 = 100$ are both greater than five).

The mean of the normal distribution is $\mu = np = 100$ and the standard deviation of the normal distribution is $\sigma = \sqrt{np(1-p)} = 7.071068$. Using the continuity correction the probability we seek is $P(x > 119.5)$, which is found by typing
$$1 - CDF.NORMAL(119.5, 100, 7.071068)$$
into the **Numeric Expression** box of the **Compute Variable** dialog box. The probability is found to be *0.002910*.

The exact binomial probability of 120 or more men is found by typing
$$1 - CDF.BINOM(119, 200, 0.5)$$
into the **Numeric Expression** box of the **Compute Variable** dialog box. The probability is found to be *0.002843*. The normal probability differs from the binomial probability by only 0.000068. The normal probability approximation to the exact binomial probability is very good.

Section 6-4   Generating a Sample of Normal Data

Sometimes it is required to simulate a normally distributed random variable with a specific mean and standard deviation. The function **RV.NORMAL(mean, stddev)** generates random values from a normal distribution with the specified mean and standard deviation. This function is similar to RV.BINOM that we saw in Chapter 4 of this manual.

Women's heights are normally distributed with a mean of *63.6* inches and a standard deviation of *2.5* inches (according to a National Health Survey). Use SPSS to simulate *1000* women's heights. In the next section, we will use these simulated heights to demonstrate how to determine if the sample came from a normal distribution.

First, we create a variable named ***heights*** with *1000* cases. Open a new data file (choose **File > New > Data**) and in the Variable View, create a new variable ***heights*** with data type *Numeric8.2*. Open the syntax editor (choose **File > New > Syntax**) and type in the code shown in Figure 6 – 1 and then click **Run > All**.

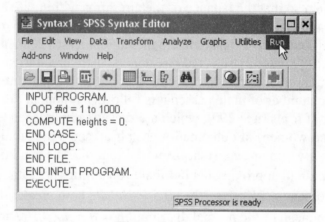

Figure 6 - 1

This will generate *1000* cases with ease case of ***heights*** set to an arbitrary value of zero.

Next, choose **Transform > Compute…** to open the **Compute Variable** dialog box. In the **Numeric Expression** box, type *RV.NORMAL(63.6, 2.5)* and type ***heights*** into the **Target Variable** box. Click the **OK** button and the random sample will appear in the variable named ***heights***.

Section 6-5   Determining Normality*

Many of the methods and procedures in statistics require that the sample data were randomly selected from a population that has a normal distribution. It is therefore necessary to determine whether sample data appear to come from a population that is normally distributed. There are several ways to check if a data set follows a normal distribution. As a first check, construct a histogram and see if the histogram *dramatically* differs from a bell-shaped symmetric curve. Make a boxplot and check that less than 1% of the observations are outliers. Finally, construct a normal quantile plot (known in SPSS as a Normal Q-Q plot) and check that the points lie close to a straight line indicating normality.

Lets check to see if the data in the variable *heights* created in the previous section follows a normal distribution. Choose **Analyze > Descriptive Statistics > Explore…** to open the **Explore** dialog box (see Figure 2 - 6). Paste the variable *heights* into the **Dependent List** box. Click the **Plots…** button to open the **Explore: Plots** dialog box (see Figure 2 - 7). Choose the checkboxes for **Histogram** and **Normality plots with tests**. Click the **Continue** button and then click the **OK** button. Sample descriptive statistics, a histogram, a normal quantile plot, and a boxplot among other items appear in the Output Viewer window.

The process that created the variable *heights* in the previous section was random, therefore every sample will be different, and yet every sample will have the property that the data values come from a normal probability distribution with a mean of *63.6* and a standard deviation of *2.5*. Since the data in the variable *heights* was generated randomly, your plots will differ from the ones shown below.

First, we will look at the histogram (Figure 6 - 2). Notice that the mean and standard deviation of the data are both close to the mean and standard deviation of the population. It appears that the center of the distribution (the peak) is centered near to the population mean of *63.6* and that the histogram does not dramatically differ from a symmetric bell-shaped curve.

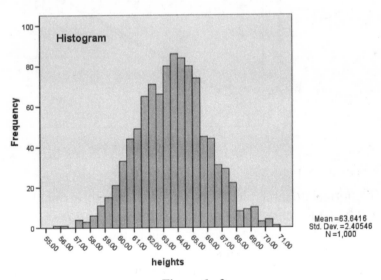

Figure 6 - 2

There is a break in the histogram between *56* and *57*, possibly indicating outliers. The boxplot (Figure 6 - 3) shows a few outliers. Since this is a very large sample (n = 1000) then we would expect about 10 outliers. This is not conclusive evidence for or against claiming the sample data in the variable *heights* came from a normal distribution.

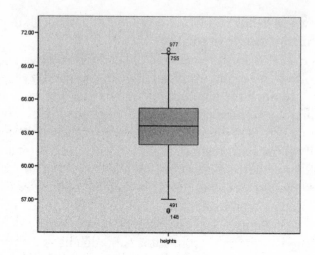

Figure 6 - 3

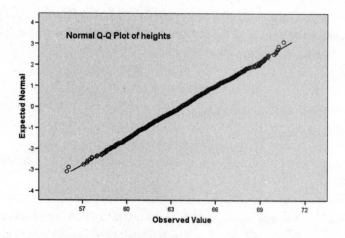

Figure 6 – 4

The normal quantile plot (Figure 6 - 4), labeled Normal Q-Q Plot of HEIGHTS in the Output Viewer window shows that the plotted points closely follow a straight line. We can finally conclude that the data in the variable *heights* came from a normal distribution. The points being plotted on the vertical axis are the *z* scores of a standard normal distribution. You can see that most of the *z* score (Expected Normal) values are less than *3* in absolute value. We can infer from this that none of the data values that occurred was extremely unlikely for a normal distribution with mean *63.6* and standard deviation *2.5*.

Section 6-6   Exercises

1. Assume that women's weights are normally distributed with a mean of *143* pounds and a standard deviation of *29* pounds. Find the following probabilities using SPSS.

   a. What is the probability that a randomly chosen woman would have a weight less than *120* pounds?
   b. What proportion of women weigh more than *135* pounds?
   c. What proportion of women weigh between *128* and *152* pounds?

2. Assume that women's weights are normally distributed with a mean of *143* pounds and a standard deviation of *29* pounds. Find the following scores using SPSS.

   a. What weight is the 90th Percentile of women's weights? That is, find the score P_{90} that separates the bottom *90%* of women's weights from the top *10%*.
   b. What is the median women's weight? That is, find the score P_{50} that separates the bottom *50%* of women's weights from the top *50%*. What is significant about this number?

3. Generate a sample of size *100* from a normal distribution with a mean of *123* and a standard deviation of *4*. What is the mean and standard deviation of the sample? Generate a second sample of size *100* from the same normal distribution. What is the mean and standard deviation of this second sample? What can you say about the means and standard deviations of random samples of the same size taken from the same population?

4. When Mendel conducted his famous hybridization experiments, he used peas with green pods and yellow pods. One experiment involved crossing peas in such a way that *25%* (or *145*) of the *580* offspring peas were expected to have yellow pods. Instead of getting *145* peas, as expected, with yellow pods, he got *152*. Assuming that Mendel's *25%* rate is correct, estimate the probability of getting at least *152* peas with yellow pods among the *580* offspring peas using the normal distribution approximation. Then calculate the exact probability using the binomial distribution. Compare the two probabilities.

5. Use the data listed in Data Set 12: **Weights and Volumes of Cola** in Appendix B of your textbook (this data is saved on the data disk as **Cola.sav**). The data set contains the weight and volume measurements on *36* cans of Regular Cola, Diet Cola, Regular Pepsi, and Diet Pepsi.

   a. Is the distribution of the weights of Regular Cola a normal distribution? Explain your answer.

b. Is the distribution of the weights of Diet Cola a normal distribution? Explain your answer.

c. Is the distribution of the weights of Regular Pepsi a normal distribution? Explain your answer.

d. Is the distribution of the weights of Diet Pepsi a normal distribution? Explain your answer.

Chapter 7

Estimates and Sample Size

Chapter 7   Estimates and Sample Sizes

Confidence intervals are one of the most common estimation techniques used in traditional statistics. Confidence intervals are interval estimates of a population parameter with an associated degree of confidence. This chapter will show how to construct confidence intervals for a population mean, a population proportion, and a population variance. The required sample size necessary to estimate a population parameter with a given margin of error will also be shown. *Elementary Statistics*, 10/e and *Essentials of Statistics*, 3/e have detailed discussions of these problems in Chapter 7. You should be familiar with Chapter 7 of your textbook prior to beginning this chapter.

Section 7-1   Estimating a Population Proportion

SPSS does not have a procedure for calculating a **confidence interval for a population proportion** or determining the **sample size to estimate a population proportion**. We can do these problems by typing the relevant formula into the **Numeric Expression** box of the **Compute Variable** dialog box.

Finding critical values associated with the Standard Normal distribution

When calculating a confidence interval by hand, as described in Chapter 7-2 of your textbook, the formulas involve finding critical values corresponding to a specified area under a standard normal curve. We will begin by showing how to determine these critical values in SPSS.

For example, when making a confidence interval with $1 - \alpha$ probability, we need to compute the critical value $z_{\alpha/2}$. This critical value has probability $1 - \alpha/2$ to the left of it and probability $\alpha/2$ to the right of it. Since the standard normal distribution is symmetric about zero, that means there is $1 - \alpha$ probability between $-z_{\alpha/2}$ and $z_{\alpha/2}$. The SPSS function **IDF.NORMAL(p, mean, stddev)** determines the critical value from a normal distribution with specified mean and standard deviation, which has probability p to the *left* of the critical value. Since $z_{\alpha/2}$ is the critical value of a standard normal distribution, which has probability $1 - \alpha/2$ to the *left* of it, we determine $z_{\alpha/2}$, using the function IDF.NORMAL($1 - \alpha/2$, 0, 1).

Find the critical value $z_{\alpha/2}$ corresponding to a *95%* confidence level. That is, find $z_{0.025}$. The critical value $z_{0.025}$ has area *0.975* to the left of it and area *0.025* to the right of it. Choose **Transform > Compute...** to open the **Compute Variable** dialog box. Type *IDF.NORMAL(0.975, 0, 1)* into the **Numeric Expression** box and name the **Target Variable** z. Click the **OK** button and SPSS calculates the critical value that has area 0.025 to its right as $z_{0.025} = 1.96$. Compare this value with one obtained from Table A-2 in Appendix A of your textbook.

Confidence Interval for a Population Proportion

Many statistical studies are concerned with obtaining the proportion (percentage) of a population that has a specified attribute. The population under consideration is usually large and it is therefore not possible to determine the population proportion by a census. Hence, a confidence interval for the population proportion is determined to give information about the parameter.

Consider the Chapter Problem *"Does Touch Therapy Work?"* at the beginning of Chapter 7 in your textbook. Emily Rosa, a fourth grade student, chose the topic of touch therapy for a science fair project. She convinced 21 experienced touch therapists to participate in a simple test of their ability to detect a human energy field. In the experiment, each therapist put both hands through holes in a cardboard partition which prevented them from seeing what was on the other side. Emily sat on the other side of the partition and placed one of her hands above the therapist's hands. Emily's choice of which hand to use was based on the toss of a single coin. The therapist was then asked to identify the hand Emily had selected. Among the 280 trials, the touch therapists identified the correct hand 123 times. We will find a **95% confidence interval for the proportion** of touch therapists who were able to identify the hand correctly.

Begin by determining the **margin of error,** $z_{0.025}\sqrt{\hat{p}\hat{q}/n}$ for this problem. Create a variable named ***phat*** that has the value *0.439286* (the proportion who made correct identifications) in the first case. Then, choose **Transform > Compute...** and type the expression

$$IDF.NORMAL(0.975, 0, 1) * SQRT(phat * (1 - phat) / 280)$$

into the **Numeric Expression** box. Name the **Target Variable** E and click the **OK** button. The resulting margin or error is *0.0581318*. The *95%* confidence interval for the population proportion is $\hat{p} - E$ to $\hat{p} + E$. Computing the confidence interval we obtain the interval *(0.381, 0.497)*.

Determining Sample Size to Estimate a Population Proportion

Chapter 7-2 of your textbook gives the formulae for determining the **sample size to estimate a population proportion** p to be either

$$n = \frac{\left[z_{\alpha/2}\right]^2 \hat{p}\hat{q}}{E^2}$$

when an estimate of $\hat{p}$ is known, or

$$n = \frac{\left[z_{\alpha/2}\right]^2 \cdot 0.25}{E^2}$$

when no estimate of $\hat{p}$ is known.

We will determine how many households must be surveyed in the following example. The use of answering machines, fax machines, voice mail, and Internet usage is growing rapidly, and they are having a dramatic effect on the way we communicate and disseminate information. Suppose a sociologist wants to determine the current percentage of U.S. households having Internet access in the home. How many households must be surveyed in order to be *95%* confident that the sample percentage is in error by no more than four percentage points?

In 2000, *41.5%* of U.S. households had Internet access (based on data from the *U.S. Census Bureau*). Using this result, we would estimate the required sample size by choosing **Transform > Compute...** and typing in the expression

(*IDF.NORMAL(0.975, 0, 1)*)**2 * 0.415 * (1 – 0.415) / 0.04** 2

into the **Numeric Expression** box. The resulting value is *582.88*, which should be increased to the next whole number *583*. That is, to be *95%* confident that the sample proportion is within *4%* of the true proportion of all households, we must randomly select a survey of *583* households.

If we had had no prior information suggesting a possible value for $\hat{p}$, we would have typed the expression

(*IDF.NORMAL(0.975, 0, 1)*)**2 * 0.25 / 0.04** 2

into the **Numeric Expression** box. The resulting value is *600.23*, which should be increased to the next whole number *601*. Therefore, when no prior information about the population proportion is available, a random sample of *601* households is required to estimate the population proportion with a margin or error of less than *4%* and *95%* confidence.

Section 7-2   Estimating a Population Mean

A **confidence interval for a population mean** can made in SPSS using the **Explore** procedure. The **Explore** dialog box (Figure 2-8) produces summary statistics, graphical displays, and a confidence interval for a population mean. As we saw in the Section 6-5 of this manual, the graphical displays allow us to check the assumptions underlying confidence intervals quickly and easily. The assumptions should always be checked prior to interpreting the confidence interval.

SPSS does not have a procedure for determining the **sample size to estimate a population mean**. However, this problem can be done, as in the proportion problem above, by typing the relevant formula into the **Numeric Expression** box in the **Compute Variable** dialog box.

Finding critical values associated with the Student t distribution

When calculating a confidence interval by hand, as described in Chapter 7-4 of your textbook, the formulas involve finding critical values corresponding to a specified area under a Student t distribution. The Student t distribution, like the standard normal distribution, is symmetric about zero. We will investigate finding critical values, denoted by $t_{\alpha/2}$. The SPSS function **IDF.T(p, df)** determines the critical value from a Student t distribution with df degrees of freedom, which has probability p to the left of the critical value. Since $t_{\alpha/2}$ is the critical value with probability $1 - \alpha/2$ to its left, then we determine $t_{\alpha/2}$ using the function IDF.T($1 - \alpha/2$, df).

Find the critical value $t_{\alpha/2}$ corresponding *13* degrees of freedom and a *95%* confidence level. That is, find $t_{0.025}$. The critical value $t_{0.025}$ has area *0.975* to the left of it and area *0.025* to the right of it. Since the Student t distribution is symmetric about zero, the critical values $-t_{0.025}$ and $t_{0.025}$ have area *0.95* between them. Choose **Transform > Compute…** to open the **Compute Variable** dialog box. Type *IDF.T(0.975, 13)* into the **Numeric Expression** box and name the **Target Variable** *t*. Click the **OK** button and SPSS calculates the critical value that has area *0.025* to its right as $t_{0.025} = 2.16$. Compare this value with one obtained from Table A-3 in Appendix A of your textbook.

Confidence Interval for a Population Mean

A **confidence interval for a population mean** μ can be made if the population being sampled is normally distributed. There is no procedure in SPSS that determines the confidence interval for a population mean when the population standard deviation is known. We will show how to use the **Explore** procedure to obtain a confidence interval for a single population mean. The **One-Sample T Test** procedure, discussed in the next chapter, can also be used to obtain a confidence interval for the population mean.

Data Set 1: **Health Exam Results** in Appendix B of your textbook (this data appears on the disk as **Fhealth.sav**) includes the pulse rates for 40 randomly selected women. Use this simple random sample of 40 data values to construct a *95%* confidence interval for μ, the mean pulse rate for all women.

Open the **Fhealth.sav** data file and choose **Analyze > Descriptive Statistics > Explore…** to open the **Explore** dialog box. In the **Explore** dialog box, paste the variable *pulse* (it is labeled *Pulse Rate (beats/min)* into the **Dependent List**. Click the **Statistics…** button to open the **Explore: Statistics** dialog box (Figure 7-1). Make sure the checkbox for **Descriptives** is selected and then type *95* into the **Confidence Interval for the Mean** box. Click the **Continue** button to close the **Explore: Statistics** dialog box.

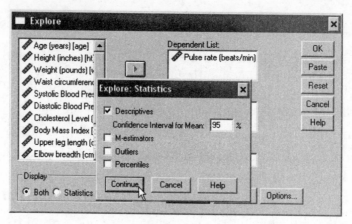

Figure 7 - 1

Then click the **OK** button to display the results in the Output Viewer window. The Descriptives table (Figure 7-2) gives the *95%* confidence interval for the mean pulse rate for all women to be *72.30* to *80.30* (rounding to two decimal places). The sample mean is found to be *76.30* and the standard error is found to be *1.98*. Other sample descriptive statistics are also shown.

Descriptives

			Statistic	Std. Error
Pulse rate (beats/min)	Mean		76.30	1.976
	95% Confidence Interval for Mean	Lower Bound	72.30	
		Upper Bound	80.30	
	5% Trimmed Mean		75.11	
	Median		74.00	
	Variance		156.215	
	Std. Deviation		12.499	
	Minimum		60	
	Maximum		124	
	Range		64	
	Interquartile Range		12	
	Skewness		1.684	.374
	Kurtosis		4.526	.733

Figure 7- 2

The Output Viewer window also includes a boxplot for these data, which can be used to check for symmetry and outliers. The boxplot shows two outliers and indicates that the distribution is approximately symmetric; therefore, we can assume that the distribution of the population is approximately normal.

Determining Sample Size to Estimate a Population Mean

Chapter 7-3 of your textbook gives the formula for determining the **sample size to estimate a population mean** to be

$$n = \left[\frac{z_{\alpha/2}\sigma}{E} \right]^2$$

where $z_{\alpha/2}$ is the critical z score based on the desired confidence level, E is the margin of error, and σ is the population standard deviation. When the population standard deviation σ is unknown, estimate the standard deviation using the rule $\sigma \approx$ range/4.

Assume we want to estimate the mean IQ for the population of statistics professors. How many statistics professors must be randomly selected for IQ tests if we want *95%* confidence that the sample mean is within *2* IQ points of the population mean? Assuming that the population standard deviation is *15* (it is known that the standard deviation of IQ tests is about *15*), the required sample size can be calculated. Choose **Transform > Compute…** and type the expression

$$(IDF.NORMAL(0.975, 0, 1) * 15 / 2)**2$$

into the **Numeric Expression** box (Figure 7-3). Name the **Target Variable *n*** and click the **OK** button. SPSS will report the necessary sample size to be *216.08*. Of course, the necessary sample size must be an integer, since we cannot sample *0.08* of a statistics professor. Hence, the required sample size is increased to the next larger whole number, namely, *217*.

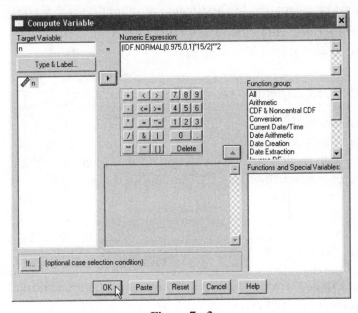

Figure 7 - 3

Section 7-3   Estimating a Population Variance

The population variance is a measure of the variation (spread) of a data set. A data set with a great deal of variation will have a large variance, whereas one with little variation will have a small variance. SPSS does not have a procedure to determine a confidence interval for a population variance but we can use the **Compute Variable** dialog box as we did with proportions.

Finding critical values associated with the Chi-square distribution

When calculating a confidence interval by hand, as described in Chapter 7-5 of your textbook, the formulas involve finding critical values corresponding to a specified area under a **chi-square distribution**. The chi-square distribution, unlike the standard normal and Student t distributions, is not a symmetric distribution. This means that you must calculate two critical values from a chi-square distribution when making a confidence interval. We will investigate finding the critical values, denoted by χ_L^2 and χ_R^2. The SPSS function **IDF.CHISQ(p, df)** determines the critical value from the chi-square distribution, with the specified degrees of freedom *df*, for which the probability to the left of the critical value is *p*.

The critical value χ_L^2 associated with *9* degrees of freedom and a *95%* confidence level has probability *0.025* to the left of it and would be calculated in SPSS using the function IDF.CHISQ(0.025, 9). The critical value χ_R^2 associated with *9* degrees of freedom and a *95%* confidence level has probability *0.975* to the left of it and would be calculated in SPSS using the function IDF.CHISQ(0.975, 9). SPSS calculates $\chi_L^2 = 2.700$ and $\chi_R^2 = 19.023$. Compare these values with the ones obtained from Table A-4 in Appendix A of your textbook.

Confidence Interval for a Population Variance

Many situations, such as quality control in a manufacturing process, require that we estimate values of population variances or standard deviations. In addition to making products with measurements yielding a desired mean, the manufacturer must make products of *consistent* quality that do not run the gamut from extremely good to extremely poor. Since this consistency can often be measured by the variance or standard deviation, these become vital statistics in maintaining the quality of products and services.

Data Set 1: **Health Exam Results** in Appendix B of your textbook (this data is saved on the data disk as **Fhealth.sav**) includes the pulse rates for 40 randomly selected women. Use the simple random sample of those *40* values to construct a *95%* confidence interval estimate for σ^2, the variance of the pulse rates for all women.

Open the **Fhealth.sav** data file. The variance of this data was found in Section 7-2 using the **Explore** procedure to be *156.215* (see Figure 7-2). The lower limit of the confidence interval is found to be *104.824* by typing

$$(40 - 1) * 156.215 / IDF.CHISQ(0.975, 39)$$

into the **Numeric Expression** box of the **Compute Variable** dialog box. Name the **Target Variable L**. The upper limit of the confidence interval is found to be *257.559* by typing

$$(40 - 1) * 156.215 / IDF.CHISQ(0.025, 39)$$

into the **Numeric Expression** box of the **Compute Variable** dialog box. Name the **Target Variable** *U*. The *95%* confidence interval for σ^2 is $L < \sigma^2 < U$ (or *104.824 < σ^2 < 257.559*). It can also be expressed as (*104.824, 257.559*).

Section 7-4   Exercises

In exercises 1-5, use the data listed in Data Set 3: **Cigarette Tar, Nicotine, and Carbon Monoxide** in Appendix B of your textbook (this data is saved on the data disk as **Cigaret.sav**). The data set contains measurements of the amount of Tar, Nicotine, and Carbon Monoxide levels of *100* mm, long, filtered cigarettes for *29* Brands.

1. If these *29* Brands represent a random sample of all cigarettes, find a *98%* confidence interval for the mean amount of Tar (measured in milligrams) found in *100* mm, long, filtered cigarettes.

2. If these *29* Brands represent a random sample of all cigarettes, find a *95%* confidence interval for the variance in the amount of Tar (measured in milligrams) found in *100* mm, long, filtered cigarettes

3. If these *29* Brands represent a random sample of all cigarettes, obtain a *95%* confidence interval for the mean amount of Carbon Monoxide found in *100* mm, long, filtered cigarettes.

4. If these *29* Brands represent a random sample of all cigarettes, obtain a *90%* confidence interval for the variance of the amount of Carbon Monoxide found in *100* mm, long, filtered cigarettes.

5. How large a random sample would be needed to estimate the mean amount of Carbon Monoxide in *100* mm, long, filtered cigarettes to within *1* milligram with *95%* confidence? Use the sample standard deviation as your estimate of σ.

6. Niki and Katie have raced against each other *47* times in the past year. Niki has won *28* of the races. Use a *96%* confidence interval to estimate the percentage of races that Niki will win.

Chapter 8

Hypothesis Testing

Chapter 8   Hypothesis Testing

A hypothesis test is a statistical procedure that is used to decide if a hypothesis (claim) about a parameter should be rejected or not, based on information in the sample data. In this chapter, we will show how to do hypothesis tests in SPSS concerning a single mean, a single proportion, or a single variance. You should be familiar with the material in Chapter 8 of *Elementary Statistics*, 10/e or *Essentials of Statistics*, 3/e prior to reading this chapter.

Section 8-1   Hypothesis Testing and P-values

A hypothesis test involves two hypotheses (claims) about a population parameter– the **null hypothesis** (H_0) and the **alternative hypothesis** (H_1). The goal is to decide which of these two hypotheses is true. That is, to decide to either reject the null hypothesis in favor of the alternative hypothesis or fail to reject the null hypothesis.

The null hypothesis is a statement that the value of the population parameter (which could be a proportion, mean, or standard deviation) is *equal* to one specific number. Examples of null hypotheses that are studied in this chapter are:

$$H_0: \; p = 0.75 \qquad H_0: \; \mu = 100 \qquad H_0: \; \sigma = 10$$

The alternative hypothesis is a statement that the parameter differs from the value stated in the null hypothesis. While the null hypothesis claims that the parameter is equal to a specific value, the alternative hypothesis claims a range of possible values for the parameter. There are three possible forms of the alternative hypothesis. Some examples of alternative hypotheses that could be associated with the null hypotheses above are:

$$H_1: \; p > 0.75 \qquad H_1: \; p < 0.75 \qquad H_1: \; p \neq 0.75$$
$$H_1: \; \mu > 100 \qquad H_1: \; \mu < 100 \qquad H_1: \; \mu \neq 100$$
$$H_1: \; \sigma > 10 \qquad H_1: \; \sigma < 10 \qquad H_1: \; \sigma \neq 10$$

Only *one* of the three forms of the alternative hypothesis will be used in a given problem. The wording of the problem tells you which form of the alternative hypothesis to use. The hypothesis tests in the first column are called **right-tailed tests**; hypothesis tests in the second column are called **left-tailed tests**; and hypothesis tests in the last column are called **two-tailed tests**. Left-tailed and right-tailed tests are **one-tailed tests**.

Recall the goal of hypothesis testing is to decide which of the two hypotheses is true. This decision is based on a **test statistic** (a number) calculated from a random sample selected from the population. *Assuming that the null hypothesis is true*, the probability of getting a value of the test statistic that is *at least as extreme* as the one representing the sample data is calculated. This probability is called the **P-value** or probability value. The P-value will then be small when the alternative hypothesis is true and large when the null hypothesis is true.

When making decisions based on test statistics, which are random variables, there is always the possibility of making an error. In fact, two different errors can be made. A **Type I error** is made if the null hypothesis is rejected when in fact it is true. A **Type II error** is made when the null hypothesis is not rejected but in fact the null hypothesis is false.

When SPSS performs any hypothesis test, it calculates a P-value. The decision criterion for a hypothesis test using the P-value is to *reject the null hypothesis* if the P-value is less than or equal to the **significance level** α. Otherwise, do not reject the null hypothesis.

> **Important note**: *SPSS always calculates the P-value of a two-tailed hypothesis test.* Consequently, if the hypothesis test being done is one-tailed the P-value must be halved (divided by 2) before comparing it to the significance level.

It is best if the significance level is selected prior to doing a hypothesis test. This is true because after calculating the P-value there may be a temptation to pick a particular significance level to obtain a desired result. Sometimes the significance level is not selected (or given) in advance. In those cases, the following rule can be used.

- *Reject the null hypothesis* if the P-value is less than *0.01*. This follows from the fact that the significance level is usually larger than *0.01*.
- *Probably reject the null hypothesis* if the P-value is between *0.01* and *0.05*. Before making your decision, select a reasonable significance level by determining the importance of making a Type I error.
- *Probably fail to reject the null hypothesis* if the P-value is between *0.05* and *0.10*. Again, before making your decision, select a reasonable significance level by determining the importance of making a Type I error.
- *Fail to Reject the null hypothesis* if the P-value is greater than *0.10*. This follows from the fact that the significance level is usually smaller than *0.10*.

Read Chapter 8-2 *Basics of Hypothesis Testing* of your textbook, for a detailed discussion of hypothesis testing and P-value.

Section 8-2   Hypothesis Testing for a Single Proportion

As we saw in Chapter 7 of this manual, SPSS does not have any procedures for calculating confidence intervals or hypothesis tests for a single proportion. We will use the **Compute Variable** dialog box to type in the relevant formulas to do the hypothesis test for a population proportion.

For example, consider the Chapter Problem *"What is the Best Way to Go About Finding a Job?"* of your textbook. Recall, based on data collected from Taylor Nelson Sofres Intersearch a recent survey involving 703 randomly selected subjects who were all working, 61% said that they found their job through a process called networking. In networking, a job seeker develops contacts and exchanges information through an informal network of people. Based on this survey, is there sufficient evidence to support the claim that a majority of workers find jobs through networking? Test this hypothesis using a *0.05* significance level.

The claim requires that we test the right-tailed test

$$H_0:\ p = 0.50 \qquad H_1:\ p > 0.50$$

Because we are testing a claim about a population proportion p, the test statistic $\hat{p}$ is relevant to this test. Create a variable named ***phat*** that has the value of *0.61* in the first case. The test statistic, z is computed by choosing **Transform > Compute...** and typing the expression

$$(\ phat - 0.50\ )\ /\ SQRT(0.5 * 0.5\ /\ 703)$$

into the **Numeric Expression** box. Name the **Target Variable** z and click the **OK** button. SPSS will report the test statistic to be *5.833*.

Next, to calculate the P-value for this test statistic, we must calculate $P(z > 5.833)$, the probability uses greater than (>) because the alternative hypothesis is right-tailed. This probability can be calculated by choosing **Transform > Compute...** and typing the expression

$$1 - CDF.NORMAL(\ z,\ 0,\ 1\ )$$

into the **Numeric Expressions** box (the variable z was calculated above). Name the **Target Variable** *pvalue* and click the **OK** button. The resulting P-value is *0.000000003*. The P-value is less than the significance level $\alpha = 0.05$ and so we reject the null hypothesis. The sample proportion of 0.61 (or 61%) falls within the range of values considered significant because they are so far above 0.5 that they are not likely to occur by chance.

Section 8-3   Hypothesis Testing for a Single Mean

Chapter 7 of your textbook discusses three different test procedures for testing a claim about a population mean. The first procedure is used when the population standard deviation is known. This rarely happens and SPSS does not have a procedure for doing this hypothesis test. When the standard deviation is unknown and the sample size is smaller than *30*, the test statistic has a Student t distribution and is called a t test. When the standard deviation is unknown and the sample size is larger than *30* (when $n > 30$), the test statistic has an approximately standard normal distribution and is called a *z* test. SPSS always does a *t* test regardless of the sample size.

The **One-Sample T Test** procedure tests whether the mean of the variable differs from a specified test value. This procedure also calculates several descriptive statistics (mean, standard deviation, and standard error) and a confidence interval for the difference between the mean and the test value.

In Section 7-2 of this manual, we found a *95%* confidence interval for μ, the mean pulse rate for women. Test the hypothesis that the mean pulse rate for women is not equal to 70 beats per minute.

In Section 7-2 of this manual, the *95%* confidence interval for the population mean was found to be *72.30 < μ < 80.30*. There is a relationship between confidence intervals and two-tailed tests. A confidence interval is an interval of values that are considered likely values for the population mean; as such, the values in the confidence interval are the same values for which we would not reject the null hypothesis (so long as the significance levels are the same). Since the claimed value *70* is not in this confidence interval, we expect to reject the null hypothesis in the following test.

The claim requires that we test the two-tailed test

$$H_0: \mu = 70 \qquad H_1: \mu \neq 70$$

Open the **Fhealth.sav** data file and choose **Analyze > Compare Means > One-Sample T Test...** to open the **One-Sample T Test** dialog box (Figure 8 - 1). Paste the variable ***pulse***, labeled *Pulse rate (beats/min)*, into the **Test Variable(s)** box. Type *70* into the **Test Value** box.

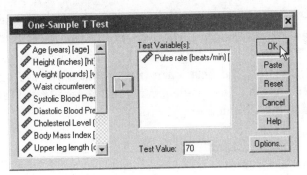

Figure 8 - 1

Click the **OK** button and the results of the **One-Sample T Test** procedure (Figure 8 - 2) will appear in the Output Viewer window. The One-Sample Statistics table in Figure 8 - 2 gives descriptive statistics for the data file. For example, the mean pulse rate for women is *76.30*. The table also gives the standard deviation and standard error of the mean.

One-Sample Statistics

	N	Mean	Std. Deviation	Std. Error Mean
Pulse rate (beats/min)	40	76.30	12.499	1.976

One-Sample Test

	Test Value = 70					
				Mean Difference	95% Confidence Interval of the Difference	
	t	df	Sig. (2-tailed)		Lower	Upper
Pulse rate (beats/min)	3.188	39	.003	6.300	2.30	10.30

Figure 8 - 2

The One-Sample Test table gives information about the **One-Sample T Test**. The test statistic was found to be *3.188*. The P-value, SPSS labels the P-value as Sig. (2-tailed), for this hypothesis test is *0.003*. Based on the P-value (recall, we reject the null hypothesis if the P-value is less than the significance level), we reject the null hypothesis and conclude that the mean pulse rate for women is not equal to *70*.

As a second example, create a data file for Problem 26 of Chapter 8-5 of your textbook. Assign the name *sugar* to the variable and set the data format to *Numeric8.2*. Save the file with the name **Sugar in Cereal**. This data file lists the sugar content (grams of sugar per gram of cereal) for a sample of *16* different cereals. Use a *0.05* significance level to test the claim of a cereal lobbyist that the mean for all cereals is less than *0.3* grams.

The claim requires that we test the one-tailed (left-tailed) test

$$H_0:\ \mu = 0.3 \qquad H_1:\ \mu < 0.3$$

Choose **Analyze > Compare Means > One-Sample T Test…** to open the **One-Sample T Test** dialog box (Figure 8 - 1). Paste the variable *sugar*, labeled *Grams of sugar per gram*, into the **Test Variable(s)** box. Type *0.3* into the **Test Value** box and click the **OK** button. The results of the **One-Sample T Test** procedure (Figure 8 - 3) will appear in the Output Viewer window.

One-Sample Statistics

	N	Mean	Std. Deviation	Std. Error Mean
sugar	16	.2950	.16773	.04193

One-Sample Test

	Test Value = 0.3					
				Mean Difference	95% Confidence Interval of the Difference	
	t	df	Sig. (2-tailed)		Lower	Upper
sugar	-.119	15	.907	-.00500	-.0944	.0844

Figure 8 - 3

The One-Sample Statistics table reports the sample mean to be *0.2950* and the standard error of the mean to be *0.04193*. The One-Sample Test table reports the test statistic for this hypothesis test to be *-0.119*. The hypothesis test being tested in this problem requires a left-tailed hypothesis but SPSS always reports the P-value for the two-tailed hypothesis test and does not have an option for changing it to reflect a one-tailed test. This is not a problem since the P-value for a one-tailed test is simply one-half the P-value of the associated two-tailed test. The P-value for the hypothesis test being considered here is then *0.4535 (0.907 / 2)*. Since the P-value is greater than the significance level $\alpha = 0.05$, we fail to reject the null hypothesis. Therefore, we cannot conclude that the mean for all cereals is less than *0.3* grams. As a result, the claim of the cereal lobbyist is not supported by the data.

Section 8-4   Hypothesis Testing for a Single Variation

The standard deviation is a measure of the variation (spread) of a data set. SPSS does not have a procedure to perform a hypothesis test for a population standard deviation. We will show how to use the **Compute Variable** dialog box to do a hypothesis test about a single standard deviation by doing Problem 15 of Chapter 8-6 of your textbook. Use a *0.01* significance level to test the claim that weights of female supermodels vary less than the weights of women in general. The standard deviation of weights of the population of women is *29* pounds. Listed below are the weights (in pounds) of nine randomly selected supermodels.

125 (Taylor)	119 (Auermann)	128 (Schiffer)
128 (MacPherson)	119 (Turlington)	127 (Hall)
105 (Moss)	123 (Mazza)	115 (Hume)

The claim requires that we test the one-tailed (left-tailed) hypotheses

$$H_0: \ \sigma = 29 \qquad H_1: \ \sigma < 29$$

Create a new data file (choose **File > Data > New**) with these *9* weights in a variable named *weights*. Then, compute descriptive statistics for this data using either the **Explore** procedure or the **Frequencies** procedure (see Chapter 3 of this manual). The standard deviation of the nine data values is *7.533*. Next, calculate the test statistic by choosing **Transform > Compute...** and typing the expression

$$(\, 9 - 1 \,) * 7.533**2 \, / 29**2$$

into the **Numeric Expression** box. Name the **Target Variable** *chisq* and click the **OK** button. SPSS reports the test statistic to be *0.5398*. The P-value of this test statistic, $P(\chi^2 < 0.5575)$ can now be computed by typing the expression (see Section 7-3 of this manual)

$$CDF.CHISQ(chisq, 8)$$

into the **Numeric Expression** box (the variable *chisq* was computed above). Name the **Target Variable** *pvalue* and click the **OK** button. SPSS computes the P-value to be *0.0002*. Therefore, we reject the null hypothesis and conclude that the variation in the weights of supermodels is less than the variation in the weights of women in general.

Section 8-5   Exercises

1. Mars, Inc., the manufacturers of M&Ms, reports on their website that *10%* of all plain-color M&Ms are blue. A recent bag of M&Ms yielded *14* blue M&Ms out of a total of *59* M&Ms in the bag. Assuming that the M&Ms in this bag are a random sample of all M&Ms, test the hypothesis, at the *0.05* significance level, that the proportion of blue M&Ms is larger than the *10%* reported by Mars, Inc.

2. Use the data shown below for the weights of a sample of gummy bears to test the claim that the mean weight of gummy bear candies is at least *0.9085* grams. Use a *0.10* significance level.

Weight	.838	.875	.870	.956	.968

3. In a math class to estimate the value of π, the diameter and circumference of seven circles of differing sizes were measured in millimeters. The resulting values for the diameter and circumference are given in the table below.

Circumference	16	22	28	35	51	75	123
Diameter	5	7	9	11	16	24	39

    Estimates of π can be obtained by finding the ratio of circumference to diameter. Enter this data into SPSS and compute the variable *ratio* (circumference divided by diameter). Based on these seven data points determine, at the *0.05* significance level, whether or not to reject the claim that the mean of the variable *ratio* is π (use $\pi = 3.14$). What is the P-value of the test?

4. The Carolina Tobacco Company advertised that its best-selling nonfiltered cigarettes contain at most *40* mg of nicotine, but *Consumer Advocate* magazine ran tests on *10* randomly selected cigarettes and found the amounts (in mg) shown in the accompanying list. It's a serious matter to charge that the company advertising is wrong, so the magazine editor chooses a significance level of $\alpha = 0.01$ in testing her belief that the mean nicotine content is greater than *40* mg. Using a *0.01* significance level, test the editor's belief that the mean is greater than *40* mg.

    47.3   39.3   40.3   38.3   46.3   43.3   42.3   49.3   40.3   46.3

5. The Newport Bottling Company has been manufacturing cans of cola with amounts having a standard deviation of 0.051 oz. A new bottling machine is tested, and a simple random sample of 24 cans results in the amounts listed in the table below. Use a 0.05 significance level to test the claim that cans of cola from the new machine have amounts with a standard deviation that is less than 0.051 oz.

11.98	11.98	11.99	11.98	11.90	12.02	11.99	11.93
12.02	12.02	12.02	11.98	12.01	12.00	11.99	11.95
11.95	11.96	11.96	12.02	11.99	12.07	11.93	12.05

Chapter 9

Inferences from Two Samples

Chapter 9   Inference from Two Samples

The previous two chapters introduced the two main procedures in inferential statistics–confidence interval estimation and hypothesis testing. Chapter 7 showed how to construct a confidence interval for one mean, one proportion, and one variance. Chapter 8 explained how to determine a hypothesis test for a single mean, a single proportion, or a single standard deviation. The goal in applied statistics is often to compare two population parameters. This chapter discusses inferential statistics based on two samples. We will learn how to calculate confidence intervals and how to determine hypothesis tests for the difference in two population proportions or two population means. We will also discuss how to compare two variances. You should be familiar with Chapter 9 of *Elementary Statistics*, 10/e or *Essentials of Statistics*, 3/e prior to beginning this chapter.

Section 9-1   Inferences about Two Proportions

As was noted in previous chapters, SPSS does not have a procedure for calculating confidence intervals or hypothesis tests for proportions. We will use the **Compute Variable** dialog box to do the calculations for hypothesis testing and confidence intervals. The hypothesis test comparing two proportions is a special case of the **Test of homogeneity** that will be discussed in Section 11-2 of this manual.

Hypothesis Test for the Difference in Two Population Proportions

Table 9 – 1 gives the number of successful treatments of CTS (carpal tunnel syndrome) for a sample of patients treated surgically and with a splint. Test the claim that treatment with surgery resulted in better outcomes than treatment with wrist splinting; use a 0.05 significance level.

	Treatment	
	Surgery	Splint
	$n_1 = 73$	$n_2 = 83$
	$x_1 = 67$	$x_2 = 60$

Table 9 - 1

The claim requires that we test the right-tailed test

$$\text{H}_0: \ p_1 = p_2 \qquad \text{H}_1: \ p_1 > p_2$$

Create three new variables, *phat1*, *phat2*, and *pbar*. The first case of *phat1* should be 0.917808 ($x_1 / n_1 = 67 / 73$), the first case of *phat2* should be 0.722892 ($x_2 / n_2 = 60 / 83$), and the first case of *pbar* should be 0.814103 (($x_1 + x_2$) / ($n_1 + n_2$) = (67 +

60) / (73 + 83)). The test statistic *z* is computed by choosing **Transform > Compute…** and typing the expression

(phat1 − phat2) / SQRT(pbar * (1 − pbar) / 73 + pbar * (1 − pbar) / 83)

into the **Numeric Expression** box. Name the **Target Variable** *z* and click the **OK** button. SPSS will report the test statistic to be *3.122550*.

Next, we calculate the P-value for this test, P(*z* > *3.122550*). This probability can be calculated by choosing **Transform > Compute…** and typing the expression

1 − CDF.NORMAL(z, 0, 1)

into the **Numeric Expression** box (the variable *z* was calculated above). Name the **Target Variable** *pvalue* and click the **OK** button. The resulting P-value is *0.000896*. Because the P-value is less than the significance level we reject the null hypothesis and conclude that there is sufficient evidence to support the claim that the proportion of successes with surgery is greater than that for splinting.

Confidence Interval for the Difference in Two Population Proportions

Use the sample data given in Table 9 - 1 to construct a *90%* confidence interval for the difference between two population proportions. If the two variables, ***phat1*** and ***phat2***, created in the example above are not in your data file then create them. We begin by determining the **margin of error** for this problem. Choose **Transform > Compute…** and type the expression

IDF.NORMAL(0.95, 0, 1) * SQRT(phat1 * (1 − phat1) / 73 + phat2 * (1 − phat2) / 83)

into the **Numeric Expression** box. Name the **Target Variable** *E* and click the **OK** button. The resulting margin of error is *0.096569*. The *90%* confidence interval for the difference in the proportion of CTS patients treated with surgery and those treated by wrist splinting is calculated by the expression (***phat1 − phat2***) ± *E* and given by the interval (*0.0983*, *0.2915*).

Section 9-2   Inferences about Two Means: Independent Samples

Depending on the assumptions there are two different two-sample independent t-tests. The first test discussed in your textbook is used when the two population variances are not equal. The second test applies only when the two population variances are equal. The **Independent-Samples T Test** procedure automatically does a confidence interval and hypothesis test for both cases. To help decide which test statistic is appropriate, SPSS provides a test for equality of variances (see Chapter 9-5 of your textbook or Section 9-4 of this manual for a discussion of this test).

The **Independent-Samples T Test** procedure compares the means for two samples. SPSS requires the data from both samples to be placed into one variable. A second variable has values that are used to separate the cases into groups.

　　To decide if the systolic blood pressure of males differs from that of females, the systolic blood pressure of five male and three female, randomly chosen subjects was measured. The data for these eight subjects are given in Table 9 - 2.

Blood Pressure	125	132	110	141	115	120	98	111
Gender	M	M	M	M	M	F	F	F

Table 9 - 2

Test the hypothesis that

$$H_0: \ \mu_{Male} = \mu_{Female} \qquad H_1: \ \mu_{Male} \neq \mu_{Female}$$

at the *0.05* significance level.

　　In a new data file, create two new variables, *bp* and *gender*. The variable *gender* should have data type to *String2*, since it will contain character data (see Section 0-3 of this manual). Type the *8* values for blood pressure into the variable *bp*. Enter the *8* data values (the M's and F's) into the variable *gender*. Value labels are optional, but the results will be more informative if you make **Value Labels** for the variable *gender* (see Section 0-3 of this manual). Because the variable *gender* is non-numeric data it cannot be used in calculations but it can be used as a classification or grouping variable.

　　To test the hypothesis that the mean male systolic blood pressure and mean female systolic blood pressure are different, choose **Analyze > Compare Means > Independent-Samples T Test...** to open the **Independent-Samples T Test** dialog box (Figure 9 - 1). Paste the variable *bp* into the **Test Variable(s)** box and the variable *gender* into the **Grouping Variable** box. After pasting *gender* into the **Grouping Variable** box it will read as *gender(? ?)*. Next, we define the two groups that correspond to the two means being compared.

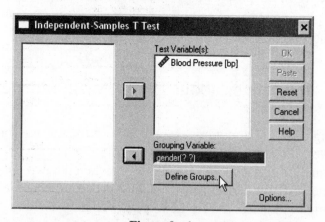

Figure 9 - 1

　　Click the **Define Groups...** button to open the **Define Groups** dialog box (Figure 9 - 2). Enter the values of the variable *gender* that are associated with **Group 1** and **Group 2**, M and F, respectively. Any case that has a value that is different from one of these two will be excluded from the analysis. Click the **Continue** button to return to the **Independent-Samples T Test** dialog box. The

Independent-Samples T Test dialog box now shows *gender('M' 'F')* in the **Grouping Variable** box; this reflects which groups were chosen for the analysis.

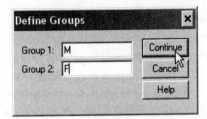

Figure 9 - 2

If we reject the hypothesis test, then the most likely next step would be to estimate the difference between male and female systolic blood pressures. Click the **Options…** button in Figure 9 - 1 to open the **Independent-Samples T Test: Options** dialog box. This dialog box is used to set the confidence level of the confidence interval for the difference in the two means. Type the desired level of confidence, say *90*, into the **Confidence Interval** box, and then click the **Continue** button to close the dialog box.

SPSS is now setup to do the calculations for a *90%* confidence interval and a hypothesis test for the difference in two means. Click the **OK** button and the output for the hypothesis test will appear in the Output Viewer window.

The Group Statistics table (Figure 9 - 3) gives some descriptive statistics for the two samples. The Independent Samples Test table reports the results of the hypothesis test that the mean blood pressure for males is different from the mean blood pressure for females.

Group Statistics

	Gender	N	Mean	Std. Deviation	Std. Error Mean
Blood pressure	Male	5	124.6000	12.54193	5.60892
	Female	3	109.6667	11.06044	6.38575

Independent Samples Test

		Levene's Test for Equality of Variances		t-test for Equality of Means							
										90% Confidence Interval of the Difference	
		F	Sig.	t	df	Sig. (2-tailed)	Mean Difference	Std. Error Difference		Lower	Upper
Blood pressure	Equal variances assumed	.182	.685	1.694	6	.141	14.9333	8.81346		-2.19280	32.05947
	Equal variances not assumed			1.757	4.837	.141	14.9333	8.49928		-2.32151	32.18818

Figure 9 - 3

Begin by deciding if the populations have equal or unequal variance. SPSS reports the P-value associated with Levene's Test for Equality of Variances (an equivalent but different test than the test discussed in Chapter 9-5 of your textbook) to be *0.685*. Levene's test is testing the null hypothesis that the variances of the two samples are equal. A significance level was not given in this problem, but the

P-value is greater than *0.10* therefore we fail to reject the null hypothesis and conclude that the two populations have equal variance.

The appropriate test statistic (the one associated with equal variances) to test the hypotheses that there is no difference in male and female blood pressures is $t = 1.694$. The P-value for this test is *0.141*. Since the P-value is greater than the significance level (*0.05*), we fail to reject the null hypothesis and conclude that there is insufficient data to establish that the mean blood pressures of men and women are different.

Notice that the test statistic associated with unequal variances, $t = 1.757$ has the same P-value as the equal variances test, *0.141*. It is unlikely for the two tests to arrive at the different conclusions when Levene's test for equality of variances is not rejected. See the discussion in Chapter 9-3 of your textbook for more in depth discussion about choosing between the equal variance and unequal variances cases. Figure 9 - 3 also provides a *90%* confidence interval for the mean difference in blood pressure between men and women to be (*-2.19, 32.06*).

As a second example, consider Data Set 13: **Weights of a Sample of M&M Plain Candies** in Appendix B of your textbook (this data is saved on the data disk as **M&M.sav**). The data set lists the weights in grams of red, orange, yellow, brown, blue, and green M&Ms found in a sample of *100* M&Ms. Test the hypothesis that the mean weight of red M&Ms is more than the mean weight of blue M&Ms at the *0.10* significance level. Also, estimate the mean difference in weights with a *98%* confidence interval. The claim requires that we test the one-tailed test

$$H_0: \mu_{Red} = \mu_{Blue} \quad H_1: \mu_{Red} > \mu_{Blue}$$

at the *0.05* significance level.

Open the **M&M.sav** data file. We will be required to specify which values associated with the value labels of red and blue M&Ms in the **Grouping Variable** dialog box. To determine which values (numbers) are associated with red and blue M&Ms, open the **Value Labels** dialog box (Figure 0-8) from the Variable View (see Section 0-3 of this manual for details). We can see that *1* is associated with the value label for red and *5* is associated with the value label for blue.

Choose **Analyze > Compare Means > Independent-Samples T Test...** to open the **Independent-Samples T Test** dialog box (Figure 9 - 1). Paste the variable *weight* into the **Test Variable(s)** box and the variable *color* into the **Grouping Variable** box. Click the **Define Groups...** button to open the **Define Groups** dialog box (Figure 9 - 2). Type *1*, the value label associated with red M&Ms, into the **Group 1** box. Then type *5*, the value label associated with blue M&Ms, into the **Group 2** box. Click the **Continue** button to return to the **Independent Sample T Test** dialog box.

Click the **Options...** button and type *98* into the **Confidence Interval** box. Click the **Continue** button and then the **OK** button. The results of the hypothesis test and confidence interval will appear in the Output Viewer window (Figure 9 - 4).

Group Statistics

	Color	N	Mean	Std. Deviation	Std. Error Mean
Weight (grams)	Red	13	.86354	.057594	.015974
	Blue	27	.85604	.041995	.008082

Independent Samples Test

		Levene's Test for Equality of Variances		t-test for Equality of Means					98% Confidence Interval of the Difference	
		F	Sig.	t	df	Sig. (2-tailed)	Mean Difference	Std. Error Difference	Lower	Upper
Weight (grams)	Equal variances assumed	.793	.379	.468	38	.642	.007501	.016028	-.031423	.046426
	Equal variances not assumed			.419	18.374	.680	.007501	.017902	-.038102	.053104

Figure 9 - 4

SPSS reports the P-value associated with Levene's Test for Equality of Variances to be *0.379*. The P-value is greater than *0.01,* therefore we fail to reject the null hypothesis and conclude that the two populations have equal variance.

The test statistic for the hypothesis that the mean weight of red M&Ms is the same as the mean weight of blue M&Ms is seen to be $t = 0.468$, with an associated degrees of freedom of *38*. The P-value that SPSS reports (*0.642*) is *two-tailed*; since we are testing a *one-tailed* test, we compute the P-value as *0.321 (0.642 / 2)*, half the P-value reported by SPSS. The P-value is greater than the significance level (*0.10*); therefore, we fail to reject the null hypothesis and conclude that there is insufficient data to prove that red M&Ms are heavier than blue M&Ms.

The *98%* confidence interval for the mean difference in the weight of red and blue M&Ms is seen to be (*-0.031, 0.046*). This confidence interval includes zero, which is consistent with the hypothesis test.

Section 9-3   Inferences from Matched Pairs

Up to this point, the methods we have studied for comparing the means of two populations have been based on two *independent* samples. We now will examine methods based on *paired* samples. A paired sample may be appropriate when there is a natural pairing of the members of the two populations. Each item in the sample consists of a *matched pair* of numbers, one from each population.

A **matched pairs** *t*-test is a one-sample *t*-test (see Section 8-3 of this manual) applied to the differences *d* for the paired data. To do a matched pairs *t*-test, you could calculate the differences by using **Transform > Compute...** procedure. Once the differences are saved in a variable, you proceed by choosing **Analyze > Compare Means > One-Sample T test...** as was done in Section 8-3. SPSS simplifies this process with the **Analyze > Compare Means > Paired-Samples T Test...** procedure.

Consider the data listed below in Table 9 - 3. This data lists the reported height and measured height of *11* male students in a Statistics class. Use the sample data to test the claim that male statistics students exaggerate their height by reporting heights that are greater than their actual height. Use a *0.05* significance level.

Student	A	B	C	D	E	F	G	H	I	J	K
Reported height	68	74	66.5	69	68	71	70	70	67	68	70
Measured height	66.8	73.9	66.1	67.2	67.9	69.4	69.9	68.6	67.9	67.6	68.8
Difference	1.2	0.1	0.4	1.8	0.1	1.6	0.1	1.4	-0.9	0.4	1.2

Table 9 - 3

The claim requires the we test the one-tailed test

$$H_0: \ \mu_d = 0 \qquad H_1: \ \mu_d > 0$$

where μ_d is the mean of the differences between reported height and measured height.

In a new data file, create two new variables, *reported* and *measured*. Type the *11 pairs* of data values into their respective variables. Save this data as the data file **Height.sav**. The data file will be used again in Chapter 12 of this manual.

Choose **Analyze > Compare Means > Paired-Samples T test…** to open the **Paired-Samples T Test** dialog box (Figure 9 - 5). First, click on the variable *reported* and SPSS will display the **Current Selection** of Variable 1 as *reported*. Next, click on the variable *measured* and SPSS will display the **Current Selections** of Variable 2 as *measured*. When both variables have been selected, click the **variable paste** ▶ button and SPSS will create the difference *reported - measured* in the **Paired Variables** box. Notice we do not calculate the differences in Table 9 - 3.

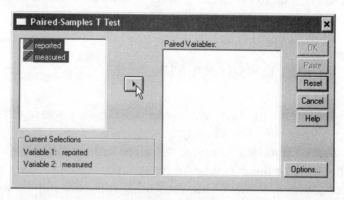

Figure 9 - 5

The **variable paste** button will not be available (it is grayed out) until two variables have been selected. The order in which you select the variables is important. SPSS computes the difference Variable 1 minus Variable 2. If you make a mistake, simply click on the difference and then click the **variable paste** button to remove the difference and start again.

The **Paired-Samples T Test** procedure also constructs a confidence interval for the mean difference. Click the **Options…** button to select a confidence level for the confidence interval. The default value is *95*. Click the **Continue** button to close the dialog box.

Click the **OK** button and the results of the hypothesis test (Figure 9 - 6) will appear in the Output Viewer window. The Paired Samples Statistics table displays descriptive statistics for the samples. The Paired Samples Correlation table (not shown) gives the correlation between the two samples. We will discuss correlation in the next chapter. The Paired Samples Test table reports information about the hypothesis test that was conducted.

Paired Samples Statistics

		Mean	N	Std. Deviation	Std. Error Mean
Pair 1	REPORTED	69.2273	11	2.11381	.63734
	MEASURED	68.5545	11	2.09445	.63150

Paired Samples Test

		Paired Differences							
					95% Confidence Interval of the Difference				
		Mean	Std. Deviation	Std. Error Mean	Lower	Upper	t	df	Sig. (2-tailed)
Pair 1	REPORTED - MEASURED	.6727	.82594	.24903	.1179	1.2276	2.701	10	.022

Figure 9 - 6

SPSS reports the test statistic as $t = 2.701$ with an associated degrees of freedom equal to *10*. The P-value of the one-tailed test is *0.011* (half the *two-tailed* P-value of *0.022*). The P-value is less than the significance level (*0.05*); therefore we reject the null hypothesis and conclude that the mean difference is greater than zero. That is, male statistics students exaggerate their height.

SPSS reports a *95%* confidence interval for the mean difference to be (*0.12*, *1.23*). That is, male statistics students exaggerate their height from between *0.12* inches to *1.23* inches.

Section 9-4   Comparing Variation in Two Samples

The characteristic of variation among data is extremely important. The variation in two independent random samples is compared using the ratio of the two sample standard deviations. If the *ratio* of the standard deviations is one then we may conclude that the standard deviations are equal. This is similar to inferences for two populations means, where we conclude the means are equal if the *difference* between the two sample means is zero. SPSS does not have a procedure for testing hypotheses about variation. We will use the **Compute Variable** dialog box to do the calculations for this hypothesis test.

Data Set 12: **Weights and Volumes of Cola** in Appendix B of your textbook (this data is saved on the data disk as **Cola.sav**) includes the weights in pounds of samples of regular Coke and regular Pepsi. The descriptive statistics for each sample are summarized in Table 9 - 4 (you can use the **Explore** procedure to obtain these values; see Section 2-4 of this manual).

	Regular Coke	Regular Pepsi
n	36	36
$\bar{x}$	0.816822	0.824103
s	0.0075074	0.0057011

Table 9 - 4

Use a *0.05* significance level to test the claim that the weights of regular Coke and the weights of regular Pepsi have the same variation. The claim requires the we test the two-tailed test

$$H_0: \ \sigma_1 = \sigma_2 \qquad H_1: \ \sigma_1 \neq \sigma_2$$

In a new data file, create two variables, *s1* and *s2*. Type the value *0.0075074* into the first case of *s1* and the value *0.0057011* into the first case of *s2*. Calculate the test statistic F by choosing **Transform > Compute...** and typing the expression

s1**2 / s2**2

into the **Numeric Expression** box. Name the **Target Variable** F and click the **OK** button. SPSS will report the test statistic to be *1.73*.

Next, we need to calculate the P-value of this test statistic. We will use the function **CDF.F(f, df1, df2)**, which computes the probability that a value from the F distribution, with degrees of freedom df1 and df2, will be less than f. The P-value, $P(F > 1.73)$ is calculated by choosing **Transform > Compute...** and typing the expression

1 – CDF.F(F, 35, 35)

into the **Numeric Expression** box. Name the **Target Variable** *pvalue* and click the **OK** button. The resulting P-value is *0.0540*. The P-value is greater than the significance level and so we fail to reject the null hypothesis and conclude that the variation in the two samples is the same.

Section 9-5   Exercises

1. In the 2000 football season, *247* plays were reviewed by officials using instant video replays, and *83* of them resulted in reversal of the original call. In the 2001 football season, *258* plays were reviewed and *89* of them were reversed (based on data from "Referees Turn to Video Aid More Often" by Richard Sandomir, *New York Times*).

    a. Is there a significant difference in the two reversal rates? Does it appear that the reversal rate was the same in both years? Use a *0.05* significance level.

b. Obtain a *95%* confidence interval for the difference in the reversal rates.

Exercises 2-6 use the data listed in Data Set 13: **Weights of a Sample of M&M Plain Candies** in Appendix B of your textbook (this data is saved on the data disk as **M&M.sav**). The data set lists the weights in grams of red, orange, yellow, brown, blue, and green M&Ms found in a sample of *100* M&Ms.

2. Test the hypothesis that orange and brown M&Ms weigh the same. Use a *0.05* significance level.

3. Determine a *90%* confidence interval for the mean difference in weight between orange and brown M&Ms.

4. Test the hypothesis that blue M&Ms are no heavier or no lighter than the other colors of M&Ms using a *0.05* significance level. **Hint**: create a new variable that separates the sample into two groups, blue and not blue. This probably is easiest done by choosing **Transform > Recode > Into different variables…**

5. Determine a *98%* confidence interval for the mean difference in weight between blue and not blue M&Ms.

6. Test the hypothesis that weights of orange M&Ms have the same variability as the weights of brown M&Ms. Use a *0.05* significance level.

The table below contains data collected by researchers on the numbers of hospital admissions resulting from vehicle crashes for Fridays on the 6th of a month and Fridays on the following 13th of the same month.

Friday the 6th	9	6	11	11	3	5
Friday the 13th	13	12	14	10	4	12

7. Use a 0.05 significance level to test the claim that when the 13th day of the month falls on a Friday, the numbers of hospital admissions from vehicle crashes are not affected.

8. Determine a 90% confidence interval for the mean difference in admissions between Fridays on the 6th and Fridays on the 13th.

9. Students at a college randomly selected *217* student cars and found they had ages with a mean of *7.89* years and a standard deviation of *3.67* years. They also randomly selected *152* faculty cars and found that they had ages with a mean of *5.99* years and a standard deviation of *3.65* years. Is there sufficient evidence to claim that the ages of faculty cars vary less than the ages of student cars?

Chapter 10

Correlation and Regression

Chapter 10   Correlation and Regression

This chapter discusses two important procedures for describing the relationship between two quantitative variables. In many problems, we want to know whether two or more variables are related, and if they are, how they are related. We will also want to know how strong the relationship between the variables is and how to make predictions. Correlation and regression are two commonly used methods for examining the relationship between two variables and for making predictions.

Correlation is used to determine whether two variables are linearly related to each other and to determine the strength of the relationship. **Regression** is a technique that calculates the *best* line that predicts a dependent variable from an independent variable. In both cases, the interest is to make inferences about the relationship between the pair of variables based on sample data. **Multiple regression** is a technique for finding a linear relationship between a dependent variable and two or more independent variables. Multiple regression, Section 10-4 of this manual, is not discussed in *Essentials of Statistics*, 3/e.

This chapter will begin by showing how to determine the correlation and how to calculate the simple linear regression equation between a pair of variables in SPSS. We will then learn how make inferences based the regression equation. Lastly, we will show how to compute a multiple regression equation in SPSS. You should be familiar with Chapter 10 of *Elementary Statistics*, 10/e or *Essentials of Statistics*, 3/e prior to beginning this chapter.

Section 10-1 Correlation

We often want to determine if there is a relationship between the two variables in a paired (*x, y*) sample. The **linear correlation coefficient, *r*** measures the strength of the *linear* relationship between paired *x*- and *y*-quantitative values in a paired sample. The values of the correlation coefficient range between *–1* and *+1*.

The two variables are said to be **negatively linearly correlated** when *r* is negative, **positively linearly correlated** when *r* is positive, and **uncorrelated** when *r* is zero. Correlation coefficients that are close to *–1* or *+1* indicate a strong linear relationship between the two variables. A correlation coefficient of exactly *+1* indicates that the points lie on a straight line with positive slope. A correlation coefficient of exactly *–1* indicates that the points lie on a straight line with negative slope. A correlation coefficient near zero indicates that there is no *linear* relationship between the two variables. This does **not** mean that the two variables are unrelated. For example, the variables can be perfectly related by a quadratic relationship and yet the correlation coefficient can be exactly zero. For this reason, we should always explore the relationship between the two variables using a scatterplot (see Section 2-3 of this manual for a description of how to make a scatterplot).

At a restaurant, we are sometimes confronted with determining how large of a gratuity to leave the waitress or waiter. There is likely to be a relationship between the amount of the bill and the amount of the gratuity. That is, we expect larger bills to result in larger gratuities. Correlation can be used to check if there is a relationship between the size of the bill and the size of the gratuity. In addition, if there is a relationship, we want to determine the relationship and predict how much of a gratuity should be left. This second part of the problem will be discussed in the next section of this manual.

Table 10 - 1 lists some sample data that shows the amount of a restaurant bill and the amount of the gratuity that was given for some recent meals at a restaurant. We will use this data to determine if there is a relationship between the amount of the bill and the amount of the tip (gratuity).

Bill ($)	33.46	50.68	87.92	98.84	63.60	107.34
Tip ($)	5.50	5.00	8.08	17.00	12.00	16.00

Table 10 - 1

In a new data file create two variables, **bill** and **tip**, that have data type *Dollar*7.2. Enter the six cases for each variable. To determine the correlation coefficient between **bill** and **tip**, choose **Analyze > Correlate > Bivariate…** to open the **Bivariate Correlations** dialog box (Figure 10 - 1). Paste both variables into the **Variables** box.

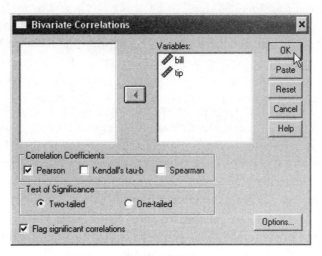

Figure 10 - 1

SPSS calculates three different correlation coefficients: Pearson, Kendall's tau-b, and Spearman. The correlation coefficient discussed in Chapter 10 of your textbook is the **Pearson's correlation coefficient**. Pearson's correlation coefficient is used when the variables being studied are quantitative, Kendall's τ_b is used for data that consist of ordered categories, and Spearman's correlation coefficient is used to measure the association between rank orders. We will discuss Spearman's correlation coefficient in Section 13-5 of this manual.

Choose the checkbox for **Pearson** because the data are quantitative. Click the **OK** button and SPSS will display the results (Figure 10 - 2) in the Output Viewer window. The correlation between the *bill* and *tip* is *0.828*. This is a large positive correlation indicating that there is a positive linear relationship between the two variables.

Correlations

		BILL	TIP
BILL	Pearson Correlation	1	.828*
	Sig. (2-tailed)	.	.042
	N	6	6
TIP	Pearson Correlation	.828*	1
	Sig. (2-tailed)	.042	.
	N	6	6

*. Correlation is significant at the 0.05 level (2-tailed).

Figure 10 - 2

Hypothesis Test concerning Correlation

To test the hypothesis that there is no linear correlation between two variables versus the alternative that there is a linear correlation, test the claim

$$H_0: \rho = 0 \qquad H_1: \rho \neq 0$$

This hypothesis test is discussed in Chapter 10-2 of your textbook. The asterisk in Figure 10 - 2 indicates that the correlation coefficient is significantly different from zero at the *0.05* significance level. SPSS reports the P-value for this hypothesis test under the estimate of the Pearson correlation coefficient to be *0.042*. Therefore, we reject the null hypothesis at the *0.05* significance level and conclude that there is a linear correlation between *bill* and *tip*. We should make a scatter diagram to see if the relationship between bill and tip is linear. See Section 2-3 of this manual for how to make the scatter diagram.

Section 10-2 Regression

Regression differs from correlation in that regression is used to predict one variable from another variable. The regression equation expresses the relationship between x (called the **independent variable, predictor variable**, or **explanatory variable**) and y (called the **dependent variable** or **response variable**). The **regression equation** is the line, $\hat{y} = b_0 + b_1 x$, that *best* describes the relationship between the dependent variable and the independent variable. It is possible to use more than one independent variable to predict the dependent variable. When more than one independent variable is used it is called multiple regression (see Section 10-4 of this manual). The case when there is only one independent variable is called **regression** (or simple linear regression).

The regression equation between *bill* and *tip* (see Table 10 - 1) is calculated by choosing **Analyze > Regression > Linear...** to open the **Linear Regression** dialog box (Figure 10 - 3). In this problem, it makes sense to predict the variable *tip* (dependent variable) from the variable *bill* (independent variable). Paste the variable *tip* into the **Dependent** box and the variable *bill* into the **Independent(s)** box.

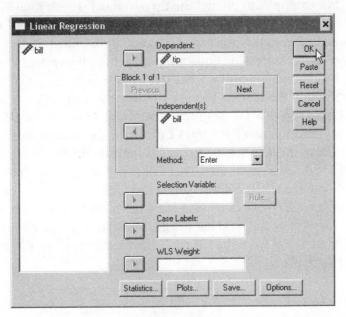

Figure 10 - 3

Click the **OK** button and several tables of regression information will appear in the Output Viewer window. We will discuss some of these tables in the next section. In this section, we are only interested in discussing the regression equation. In the next section, we will discuss measures of the quality of the regression equation.

The Coefficients table (Figure 10 - 4) shows the value of the y-intercept (b_0) and the slope (b_1) for regression equation ($\hat{y} = b_0 + b_1 x$). The value of the y-intercept is *-0.347* and the value of the slope is *0.149*. In this problem, the slope (b_1) corresponds to the amount of tip per one unit (1) change in the total bill. That is, for each additional dollar on the bill, the tip increases by *14.9¢*. The regression equation is $\hat{y} = -0.347 + 0.149\,x$. The regression equation estimates the tip to be *14.9%* of the total bill less *34.7¢*.

Coefficients[a]

Model		Unstandardized Coefficients		Standardized Coefficients	t	Sig.
		B	Std. Error	Beta		
1	(Constant)	-.347	3.936		-.088	.934
	BILL	.149	.050	.828	2.955	.042

a. Dependent Variable: TIP

Figure 10 - 4

The regression line can be used to predict the tip of a meal whose total bill is known. For example, to predict the tip of a meal whose total bill is $50.00, use

$$\hat{y} = -0.347 + 0.149\ (50.00) = 7.103 \text{ or } \$7.10.$$

It is desirable to plot the regression equation onto a scatter diagram of the data to see how well the regression equation predicts the dependent variable. Create a scatter diagram by choosing **Graphs > Scatter...** (see Section 2-3 of this manual). Select the scatterplot in the Output Viewer window by clicking on it and then choose **Edit > SPSS Chart Object > Open** (alternatively you could simply double-click on the scatter diagram) to open the **Chart Editor** for the scatterplots.

The **Chart Editor** is specific to the chart being edited (see Section 2-2 of this manual for a discussion of the Chart Editor). Most of the charts (graphs) in SPSS can be edited with the Chart Editor. There are different options for each type of chart (e.g. Scatterplots, Histograms, and Pie Charts, etc). The chart editor menus should look familiar to anyone who has reached this point in the manual. Choose **Elements > Fit Line at Total** from the **Chart Editor** menu to open the **Properities** dialog box (Figure 10 - 5).

Figure 10 - 5

In the **Properites** dialog box, click the **Fit Line** tab (Figure 10-6). Choose the button for **Linear** and click the **Close** button to close the dialog box and return to the **Chart Editor** window.

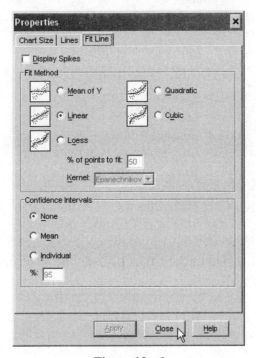

Figure 10 - 6

Choose **File > Close** from the **Chart Editor** menu to update the scatterplot in the Output Viewer window. The regression equation is now displayed on the scatterplot (Figure 10 - 7) in the Output Viewer window.

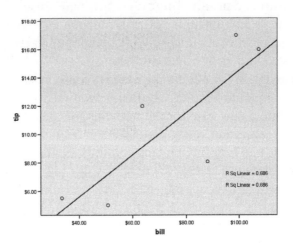

Figure 10 - 7

It can be seen that the line passes through the middle of points and that the regression equation, $\hat{y} = -0.347 + 0.149(x)$, is a good estimate of the variable *tip*.

Section 10-3   Variation and Prediction Intervals

We now turn our attention to assessing the quality of the regression equation. The coefficient of determination and the standard error of the estimate help us assess how good the regression equation is at predicting the dependent variable. The **coefficient of determination** is the amount of variation in *y* that is explained by the regression line. The **standard error of the estimate**, denoted by s_e, is a measure of the differences (or distances) between the observed sample *y*-values and the predicted values $\hat{y}$ that are obtained from the regression equation.

Another table that results from the Linear Regression procedure is the Model Summary table (Figure 10 - 8). It shows several statistics used to measure the quality of the regression equation. It can be seen that the **correlation coefficient** is *0.828*, which is the Pearson correlation coefficient found in Section 10-2 of this manual. The **coefficient of determination** is *0.686*; often the coefficient of determination is reported as a percentage (i.e. *68.6%*). This means that *68.6%* of the total variation in the variable *tip* is explained by the regression line. The **standard error of the estimate** is seen to be *$3.27* (rounded to two decimal places).

Model Summary

Model	R	R Square	Adjusted R Square	Std. Error of the Estimate
1	.828ᵃ	.686	.607	$3.26581

a. Predictors: (Constant), BILL

Figure 10 - 8

Prediction Intervals

SPSS can construct a **prediction interval**, which is a confidence interval for the predicted value $\hat{y}$. For example, we will obtain a *90%* prediction interval for the gratuity of a meal whose bill was *$75.00*. Computing the prediction interval by hand requires the standard error of the estimate, s_e.

We are to obtain a prediction for the response variable *tip* when the predictor variable *bill* is equal to *$75.00*. To do this we first add a new case to the data file by typing *75.00* into a new case for the variable *bill*. When we do this, SPSS will give the variable *tip* a missing value, denoted by a dot (.) in the case. Since the new case has a missing value, it will not be used in the calculations for the regression line. If we wanted prediction intervals for several values of the variable *bill*, we would add several new cases to the data file. SPSS will construct a prediction interval for each case in the data file. Next, click the **Save** button on the **Linear Regression** dialog box (Figure 10 - 3) to open the **Linear Regression: Save** dialog box (Figure 10 - 9).

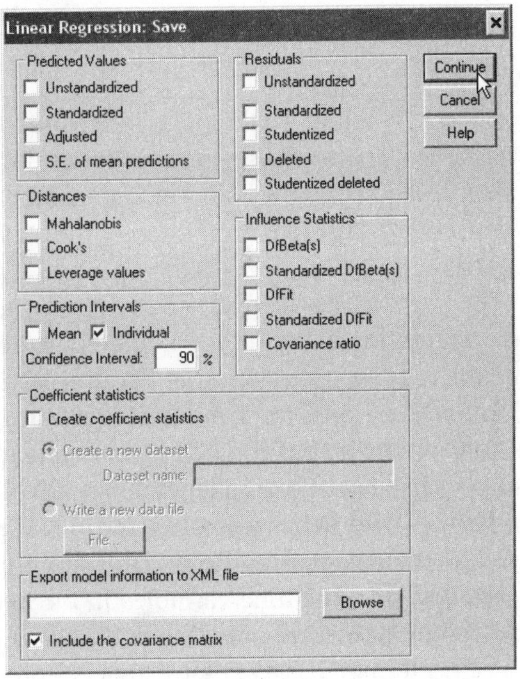

Figure 10 - 9

This dialog box can be used to save many different numbers relating to the regression line. For example, this dialog box can be used to save predicted values, residuals, and other statistics useful for diagnostics. Each selection adds one or more new variables to the data file. Choose the checkbox for **Individual** in the **Prediction Intervals** area. Enter the required confidence, *90*, into the **Confidence Interval** box. Click the **Continue** button to close the dialog box and the **OK** button to perform the regression.

SPSS will add two new variables ***lici_1*** and ***uici_1*** to the data file. These two variables are the lower and upper endpoints of the required prediction interval of the response for a particular value of ***bill***. Browse the Data Editor (Figure 10-10) until you find the case with ***bill*** = *$75.00*, where you will find the *90%* prediction interval for the gratuity is $*3.28* to $*18.32* (rounded to two decimal places).

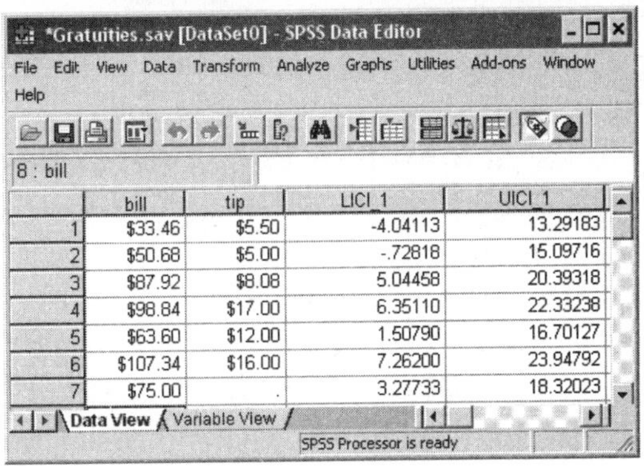

	bill	tip	LICI_1	UICI_1
1	$33.46	$5.50	-4.04113	13.29183
2	$50.68	$5.00	-.72818	15.09716
3	$87.92	$8.08	5.04458	20.39318
4	$98.84	$17.00	6.35110	22.33238
5	$63.60	$12.00	1.50790	16.70127
6	$107.34	$16.00	7.26200	23.94792
7	$75.00	.	3.27733	18.32023

Figure 10 - 10

Section 10-4 Multiple Regression*

Multiple regression is when more than one independent variable is used to predict the dependent variable. The **multiple regression equation** is the line that *best* describes the linear relationship between the dependent variable and two or more (say *k*) independent variables. Multiple regression is the statistical technique that estimates the coefficients b_0, b_1, ... b_k in the multiple regression line

$$\hat{y} = b_0 + b_1 x_1 + b_2 x_2 + ... + b_k x_k.$$

The multiple regression equation is calculated using the same procedure that was used to calculate the regression equation in Section 10-2. Choose **Analyze > Regression > Linear...** to open the **Linear Regression** dialog box (Figure 10 - 3). A multiple regression is done by pasting *k* independent variables into the **Independent(s)** box, instead of one independent variable.

Data Set 6: **Bears (wild bears anesthetized)** in Appendix B of your textbook (this data is saved on the data disk as **Bears.sav**) has nine measurements taken from *54* wild anesthetized bears. The nine variables in the data file are: *age* (age in months), *month* (month of measurement), *sex*, *headlen* (length of head in inches), *headwth* (width of head in inches), *neck* (distance around neck in inches), *length* (length of body in inches), *chest* (distance around the chest in inches), and *weight* (measured weight in pounds). We will find a linear equation to predict the weight of a bear from the length of head and length of body.

Open the data file **Bears.sav** and choose **Analyze > Regression > Linear...** to open the **Linear Regression** dialog box (Figure 10 - 3). Paste *weight* into the **Dependent** box, and paste the variables *headlen* and *length* into the *Independent(s)* box. Click the **OK** button and the results of the multiple regression equation will appear in the Output Viewer window.

The Coefficients table (Figure 10 - 11) gives information about the coefficients of the regression equation. It can be seen from the table that the multiple regression equation is

$$\hat{y} = -424.8 + 14.4 \ headlen + 7.2 \ length.$$

Coefficients^a

Model		Unstandardized Coefficients		Standardized Coefficients		
		B	Std. Error	Beta	t	Sig.
1	(Constant)	-424.804	51.495		-8.249	.000
	Length of Head	14.406	10.001	.254	1.440	.156
	Length of body	7.184	2.004	.631	3.585	.001

a. Dependent Variable: Measured Weight

Figure 10 - 11

The multiple regression equation can be used to predict the weight of a bear from its head length and body length. For example, the multiple regression equation predicts the weight of a bear with a head length of *12.0* inches and a body length of *72.0* inches to be

$$\hat{y} = -424.8 + 14.4\,(\,12.0\,) + 7.2\,(\,72.0\,) = 266.4\,.$$

A bear with these measurements will weigh approximately *266.4* pounds.

A *90%* prediction interval for the weight of a bear with **headlen** = *12.0* and **length** = *72.0* us done in the same way as for a simple linear regression (see Section 10-3 of this manual for details). To make *90%* prediction interval, add a new case to the data file with both of these values. Then, click the Save button to open the Linear Regression: Save dialog box (see Figure 10 - 9). Choose the checkbox for **Individual** and click the **OK** button to perform the regression and compute the prediction interval.

The Model Summary table (Figure 10 - 12) shows several statistics used to determine the quality of the multiple regression equation. It can be seen that the multiple correlation coefficient is *0.870*. This value is now a generalized version of the Pearson correlation coefficient found in Section 10-2. The **multiple coefficient of determination** measures the percentage of the total variation in the dependent variable **weight** explained by the regression equation. SPSS reports this value be *0.757* (or *75.7%*). This means that *75.7%* of the total variation in the variable **weight** was explained by the *two* variables, **headlen** and **length**. The **adjusted coefficient of determination**, named Adjusted R Square in SPSS, is a more appropriate measure of goodness for a multiple regression equation than R Square (see Chapter 10-5 of *Elementary Statistics*, 10/e for an explanation). The Adjusted R Square is *0.748* or *74.8%*. The **standard error of the estimate** is *61.20*. The standard error of the estimate is a measure of the differences between the dependent variable **weight** and the predicted values $\hat{y}$.

Model Summary

Model	R	R Square	Adjusted R Square	Std. Error of the Estimate
1	.870[a]	.757	.748	61.198

a. Predictors: (Constant), Length of body, Length of Head

Figure 10 - 12

Section 10-5 Exercises

Exercises 1-6 use the data listed in Data Set 6: **Bears (wild bears anesthetized)** in Appendix B of your textbook (this data is saved on the data disk as **Bears.sav**). The data set contains nine measurements on *54* anesthetized wild bears. The nine variables in the data file are: *age* (age in months), *month* (month of measurement), *sex*, *headlen* (length of head in inches), *headwth* (width of head in inches), *neck* (distance around neck in inches), *length* (length of body in inches), *chest* (distance around the chest in inches), and *weight* (measured weight in pounds).

1. Find the correlation between *weight* and each of the variables *age, headlen, headwth, neck, length,* and *chest*. Which variable is the most correlated with *weight*? Which variable is least correlated with *weight*?

2. Explain why it would not be appropriate to calculate the Pearson's correlation coefficient between *month* and *weight*.

3. Determine the regression equation predicting *weight* from the variable *chest*. What is the regression equation? What percent of the total variation of the variable *weight* did the variable *chest* explain?

4. Use a *98%* prediction interval to predict the weight of a bear with a chest measurement of *60* inches.

5. Determine the multiple regression equation predicting *weight* from the variables, *neck* and *chest*. What is the multiple regression equation? What is the adjusted coefficient of determination associated with this regression equation? What percent of the total variation of the variable *weight* does the multiple regression equation explaining?

6. Use a *95%* prediction interval to predict the weight of a bear with a chest measurement of *60* inches and a neck measurement of *30* inches.

Chapter 11

Multinomial Experiments and Contingency Tables

Chapter 11   Multinomial Experiments and Contingency Tables

The statistical techniques discussed thus far have dealt exclusively with inferences about population parameters. In this chapter, we will discuss some widely used inferential procedures that are not concerned with population parameters. These procedures are often referred to as chi-square procedures because they rely on the chi-square distribution. We will discuss two chi-square procedures– **Goodness-of-Fit** and **Tests of Independence**. You should read Chapter 11 of *Elementary Statistics*, 10/e or *Essentials of Statistics*, 3/e prior to reading this chapter.

A multinomial experiment is similar to a binomial experiment except that the experiment has more than two categories. A **multinomial experiment** results when the following conditions occur: the number of trials is fixed, the trials are independent, the outcomes of each trial must be classified into exactly one of several different categories, and the probabilities of the different categories remain constant. Specifically, we are interested in the case when a random sample of data has been separated into different categories. The **Goodness-of-Fit** test is used to test the hypothesis that the observed frequency distribution of the categories fits some claimed distribution. The **Chi-square Test** procedure in SPSS does a Goodness-of-Fit test.

Sometimes, a random sample is classified into a contingency table (or two-way frequency table) by two categorical variables. The question of interest, in this situation, is to determine if the row and column variables are independent (non-associated). This hypothesis test is called a **Test of Independence**. The **Crosstabs** procedure in SPSS does tests of independence.

Section 11-1 Multinomial Experiments: Goodness-of-Fit

When a random sample is classified according to some qualitative (categorical) variable into two or more (say k) categories, we use a **Goodness-of-Fit** test to test the hypothesis that the observed frequency distribution agrees with some theorized distribution. That is, the **Goodness-of-Fit** procedure tests the null and alternative hypotheses:

$$H_0: \quad p_1 = p_1^*, \qquad p_2 = p_2^*, \ldots, \qquad p_k = p_k^*$$
$$H_1: \text{at least one of these } k \text{ statements is not true}$$

where $p_1, p_2, \ldots p_k$ are the proportion of the sample classified into the kth category, respectively, and $p_1^*, p_2^*, \ldots, p_k^*$ are the theorized probabilities of the kth category, respectively.

Mars, Inc., claims that its M&M plain candies are distributed with the following color percentages: 13% red, *20%* orange, *14%* yellow, *13%* brown, 24% blue, and *16%* green. We will test this claim using Data Set 13: **Weights of a Sample of M&M Plain Candies** of your textbook (this data is saved on the data disk as **M&M.sav**). This data set lists the weights in grams of red, orange, yellow, brown, blue, and green M&Ms found in a sample of *100* M&Ms. The colors of the M&Ms listed the sample can be summarized using the **Explore** procedure (see Section 2-2 of this manual).

Table 11 - 1 lists the number of M&Ms in the sample for each color. Test the claim that the color distribution in the sample is the same as claimed by Mars, Inc.; use a *0.05* significance level.

	Red	Orange	Yellow	Brown	Blue	Green
Observed frequency	13	25	8	8	27	19

Table 11 - 1

In a new data file, create two variables, *color* and *number*. The variable *color* should take on the values *1*, *2*, *3*, *4*, *5*, and *6*, associated with the value labels: Red, Orange, Yellow, Brown, Blue, and Green, respectively. The variable *number* should take on the observed frequencies listed in the table above.

This data is **summary data** since we do not have all *100* cases in the data file but instead only have the total number in each category. When an SPSS data file contains summary data, it is necessary to **weight the cases** prior to doing any analyses. The values of the weighting variable indicate the number of observations represented by a single case in the data file. In this problem, we want to weight the cases according to the variable *number*. Choose **Data > Weight Cases...** to open the **Weight Cases** dialog box (Figure 11 - 1). Choose the bullet for **Weight cases by** and paste the variable *number* into the **Frequency Variable** box. Click the **OK** button. The cases are now weighted by the variable *number*.

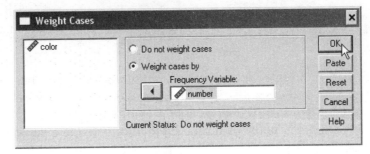

Figure 11 - 1

Once a weight variable is applied, it remains in effect until either another weight variable is applied or weighting is turned off. To turn off weighting, choose the bullet for **Do not weight cases** in the **Weight Cases** dialog box (Figure 11 - 1). The weighting information is saved along with the data values when an SPSS data file is saved. Weighting can be turned off at any time, even after the file has been saved in weighted form.

The claim requires that we test the hypothesis

H_0: $p_{Red} = 0.13$, $p_{Orange} = 0.20$, $p_{Yellow} = 0.14$, $p_{Brown} = 0.13$, $p_{Blue} = 0.24$, $p_{Green} = 0.16$

H_1: at least one of the 6 probability statements is untrue.

The values p_{Red}, p_{Orange}, p_{Yellow}, p_{Brown}, p_{Blue}, and p_{Green}, represent the proportion of the sample that are a particular color. The values 0.13, 0.20, 0.14, 0.13, 0.24, and 0.16, are the theorized probabilities for the particular color claimed by Mars, Inc. The null hypothesis consists of six probability statements. The null hypothesis will be rejected if any of these six statements is found to be incorrect.

Choose **Analyze > Nonparametric Tests > Chi-Square...** to open the **Chi-Square Test** dialog box (Figure 11 - 2). Paste the variable *color* into the **Test Variable List** box. Next, set up the **Expected Values**. The Expected Values are the hypothesized probabilities, 0.13, 0.20, 0.14, 0.13, 0.24, and 0.16, stated in the null hypothesis. The Expected Values are not equal therefore choose the bullet for **Values**.

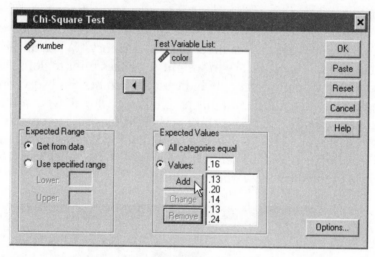

Figure 11 - 2

It is important that the values appear in the **Expected Values** box in the order in which the colors are in the data file. Each time you add a value, it appears at the bottom of the value list. The order of the values is important, since it corresponds to the ascending order of the category values of the variable *color*, in this case, red, orange, yellow, brown, blue, and green. Type the expected value (0.13) associated with the value of *color* equal to 1 (Red) into the box beside **Values**, and then click the **Add** button. Next, type the expected value (0.20) associated with the value of *color* equal to 2 (Orange), and then click the **Add** button. Continue until all 6 expected values have been entered. If you enter an expected value incorrectly, click on the incorrect value and enter the correct value into the **Values** box. Click the **Change** button to replace the incorrect value.

Click the **OK** button and the Color table (Figure 11 - 3) will be displayed in the Output Viewer window. Check to make sure that the Observed N (values of the variable ***number***) and Expected N (probabilities claimed in the null hypothesis) are correctly entered.

color

	Observed N	Expected N	Residual
Red	13	13.0	.0
Orange	25	20.0	5.0
Yellow	8	14.0	-6.0
Brown	8	13.0	-5.0
Blue	27	24.0	3.0
Green	19	16.0	3.0
Total	100		

Figure 11 - 3

The Test Statistics table (Figure 11 - 4) reports the value of the χ^2 test to be *6.682*. The P-value of the Goodness-of-Fit test is *0.245*. The P-value is greater than the significance level. Therefore, we fail to reject the null hypothesis and conclude that there is insufficient information to conclude that the percentages of M&Ms given by Mars, Inc are incorrect.

Test Statistics

	color
Chi-Square[a]	6.682
df	5
Asymp. Sig.	.245

a. 0 cells (.0%) have expected frequencies less than
5. The minimum expected cell frequency is 13.0.

Figure 11 - 4

Section 11-2 Contingency Tables: Independence and Homogeneity

In a multinomial experiment, data are classified by a single variable. For this reason, multinomial experiment data are sometimes called one-way tables. Data that are classified by observing two variables of a population are called bivariate data. A frequency distribution for bivariate data is called a **contingency table** or a **two-way table**. There are two tests included in this section: a **test of independence** between the row and column variable and a **test of homogeneity**. The test of homogeneity tests the claim that different populations have the same proportion of some characteristics.

Test of Independence

When a random sample is classified into cells by two categorical variables (a row variable and a column variable), we use a χ^2 **test of independence** to test if there is an association between the row variable and the column variable. That is, to test if the row variable and the column variable are independent or dependent. The null and alternative hypotheses are

H_0: The row and column variables are independent
H_1: The row and column variables are dependent

Table 11 - 2 summarizes the retrospective study of motorcycle drivers (this data is taken from Chapter 11-3 of your textbook). The row variable has two categories representing the *fate* of the motorcyclists: controls and cases. Subjects in the control category were motorcycle riders randomly selected at roadside locations. Subjects in the case category were motorcycle drivers seriously injured or killed. The column variable is used to *group* the motorcyclists by the color of the helmet they were wearing.

	Black	White	Yellow / Orange	Total
Controls (not injured	491	377	31	899
Cases (injured or killed	213	112	8	333
Total	**704**	**489**	**39**	1232

Table 11 - 2

In a new data file, create three variables, *group*, *fate*, and *count*. The variable *group* takes the values *1*, *2*, and *3*, associated with the value labels: Black, White, and Yellow/Orange, respectively. The variable *fate* takes the values *1* and *2*, associated with the value labels: "Not Injured" <u>and</u> "Injured or Killed", respectively. The variable *count* is the number associated with the different categories of *group* and *fate*.

The variables *group* and *fate* separate the data into *6* categories (or cases). The variable *count* is the number of people in each of these categories. Data of this type are called categorical (qualitative or attribute) data. After entering the data, the Data View should look like Figure 11 - 5.

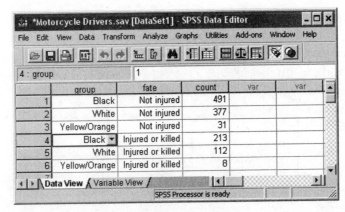

Figure 11 - 5

This data is **summary data** since we do not have all *1232* cases in the data file but instead only have the total number in each category. When an SPSS data file contains summary data, it is necessary to weight the cases **prior to doing any analyses**. The values of the weighting variable indicate the number of observations represented by a single case in the data file. To weight the cases according to the variable *count*, choose **Data > Weight Cases...** to open the **Weight Cases** dialog box (see Figure 11 - 1). Choose the bullet for **Weight cases by** and paste the variable *count* into the **Frequency Variable** box. Click the **OK** button. Cases are now weighted by *count*.

It is a good idea to check if the data is properly entered prior to doing any analyses. Choose **Analyze > Descriptive Statistics > Crosstabs...** to open the **Crosstabs** dialog box (Figure 11 - 6). Paste the variable *fate* into the **Row(s)** box and the variable *group* into the **Columns(s)** box.

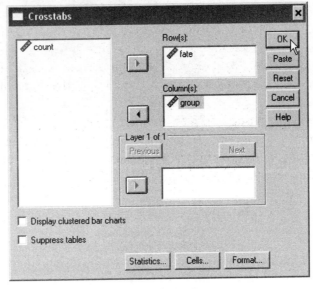

Figure 11 - 6

Click the **OK** button and the FATE * GROUP Crosstabulation table (Figure 11 - 7) will be displayed in the Output Viewer window. Compare this table with Table 11 - 2.

fate * group Crosstabulation

Count

		group			Total
		Black	White	Yellow/ Orange	
fate	Not injured	491	377	31	899
	Injured or killed	213	112	8	333
Total		704	489	39	1232

Figure 11 - 7

Use a *0.05* significance level, to test the claim that whether a motorcyclist is not injured or injured/killed is independent of whether the motorcyclist wore a black, white, or yellow/orange helmet. That is, test:

H_0: *fate* and *group* are independent
H_1: *fate* and *group* are dependent

To test this hypothesis click the **Statistics...** button on the **Crosstabs** dialog box (Figure 11 - 7) to open the **Crosstabs: Statistics** dialog box (Figure 11 - 8). This dialog box has many different statistics and measures of association from which we can choose. Choose the checkbox for **Chi-square** to test that the row variable (*fate*) and column variable (*group*) are independent. Click the **Continue** button to close the dialog box.

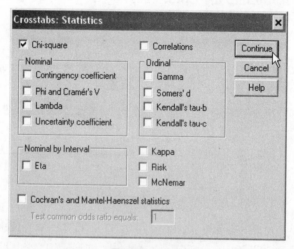

Figure 11 - 8

SPSS displays the Observed frequencies by default; the Expected frequencies can be displayed as well by clicking the **Cells...** button in the **Crosstabs** dialog box (Figure 11 - 6) to open the **Crosstabs: Cell Display** dialog box (Figure 11 - 9). This dialog box has several different statistics

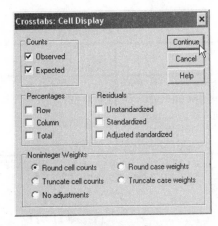

Figure 11 - 9

associated with each cell from which we can choose. Choose the checkboxes for **Observed** and **Expected** to display the observed and expected counts in each cell. The checkbox for **Observed** is probably already selected.

Click the **Continue** button and then the **OK** button to display the FATE * GROUP Crosstabulation table (Figure 11 - 10) in the Output Viewer window. The observed and expected counts are now included in the table (compare with Figure 11 - 7). The expected counts in the cells are calculated assuming that the variables *fate* and *group* are independent. Looking at the table, we can see that more motorcyclists wearing black helmets died and fewer motorcyclists died wearing white helmets than would have been expected if *fate* and *group* were independent. The numbers of motorcyclists wearing yellow/orange helmets is slightly less than would be expected if the variables were independent.

fate * group Crosstabulation

			group			
			Black	White	Yellow/Orange	Total
fate	Not injured	Count	491	377	31	899
		Expected Count	513.7	356.8	28.5	899.0
	Injured or killed	Count	213	112	8	333
		Expected Count	190.3	132.2	10.5	333.0
Total		Count	704	489	39	1232
		Expected Count	704.0	489.0	39.0	1232.0

Figure 11 - 10

The Chi-square Tests table (Figure 11 - 11) gives several test statistics: the Pearson Chi-Square, the Likelihood Ratio, and the Linear-by-Linear Association test statistics. The Pearson Chi-Square test statistic is the χ^2 test of independence discussed in Chapter 11 of your textbook. The χ^2 test statistic is associated with *2* degrees of freedom is found to be *8.775*.

Chi-Square Tests

	Value	df	Asymp. Sig. (2-sided)
Pearson Chi-Square	8.775[a]	2	.012
Likelihood Ratio	8.887	2	.012
Linear-by-Linear Association	8.414	1	.004
N of Valid Cases	1232		

a. 0 cells (.0%) have expected count less than 5. The minimum expected count is 10.54.

Figure 11 - 11

SPSS reports the P-value to be *0.012,* which is less than the *0.05* significance level. We reject the null hypothesis and conclude the variables, *fate* and *group* are not independent.

Clearly, the death rates are different for the various groups. For example, notice that *30.3% (213 / 704)* of the motorcyclists wearing a black helmet were injured or killed, *22.9% (112 / 489)* of those wearing a white helmet were injured or killed, and died, *20.5% (8 / 39)* of those wearing a yellow/orange helmet were injured or killed. This is consistent with the conclusion that the fate of a motorcyclist and whether a black, white, or yellow/orange helmet is worn are dependent variables.

Test of Homogeneity

The **test of homogeneity** is a test that different populations have the same proportions for some characteristics. In a test of homogeneity, *two or more* samples from different populations will be classified by a categorical variable and the question is if the proportion of each population in each category is the same. This is different from the *motorcycle* example in which *one* sample of data was classified by two variables.

U.S. News & World Reports claimed in an article that "On sensitive issues, people tend to give *acceptable* rather than honest responses; their answers may depend on the gender or race of the interviewer." To support that claim, an Eagleton Institute poll was conducted in which surveyed men were asked if they agreed with the statement "Abortion is a private matter that should be left to the woman to decide without government intervention." The survey was designed so that male interviewers were instructed to obtain *800* responses from male subjects and female interviewers were instructed to obtain *400* responses from male subjects. Using a *0.05* significance level, to test the claim that the proportions of agree/disagree responses are the same for the subjects interviewed by men and the subjects interviewed by women. Table 11 - 3 shows the responses of surveyed men by gender of the interviewer.

	Gender of Interviewer	
	Man	Woman
Men who agree	560	308
Men who disagree	240	92

Table 11 - 3

Because we have two different populations (subjects interviewed by men and subjects interviewed by women), this is a **test of homogeneity**. The test of homogeneity is done in exactly the same way as the test of independence. The **Crosstabs** procedure is used to do both tests. The claim requires that we test the hypothesis

H_0: The proportions of agree/disagree responses are the same for each population

H_1: The proportions are different

In a new data file, create three variables, *gender*, *response*, and *count*. The variable *gender* takes the values *1* and *2*, associated with the value labels: Man and Woman, respectively. The variable *response* takes the values *1* and *2*, associated with the value labels: Men who agree and Men who disagree, respectively. The variable *count* is the number associated with the different categories of *gender* and *response*. This is **summary data**, therefore open the **Weight Cases** dialog box (Figure 11 - 1) and weight the cases by *count* (see Section 11-1 of this manual).

Choose **Analyze > Descriptive Statistics > Crosstabs...** to open the **Crosstabs** dialog box (Figure 11 - 6). Paste the variable *response* into the **Row(s)** box and the variable *gender* into the **Columns(s)** box. Click the **Statistics...** button to open the **Crosstabs: Statistics** (Figure 11 - 8) dialog box. Choose the checkbox for **Chi-square** to test that the populations defined by the column variable (*gender*) have homogeneous responses. Click the **Continue** button to close the **Crosstabs: Statistics** dialog box.

Click the **Cells...** button to open the **Crosstabs: Cell Display** dialog box (Figure 11 - 9). Instead of choosing the checkboxes for **Observed** and **Expected**, choose the checkbox for **Column** in the **Percentages** section of the dialog box. In the test of homogeneity, we want to know if the percentage of men who agree with the statement is the same for each population. Choosing the checkbox for **Column** will display the percentage of all the cases in the column that fall into the agree and disagree cells. Uncheck the boxes for **Observed** and **Expected** as this additional information clutters the table and makes the results difficult to read.

Click the **OK** button and the REPSONSE * GENDER Crosstabulation table (Figure 11 - 12) will be displayed in the Output Viewer window. It is clear that the percentage of Men who agree with the statement is greater when the interviewer is a woman. The test of homogeneity will test whether this is a significant difference.

% within GENDER **RESPONSE * GENDER Crosstabulation**

		GENDER		Total
		Man	Woman	
RESPONSE	Men who agree	70.0%	77.0%	72.3%
	Men who disagree	30.0%	23.0%	27.7%
Total		100.0%	100.0%	100.0%

Figure 11 - 12

The Chi-Square Tests table (Figure 11 - 13) displays several statistics. The Pearson Chi-Square value is the test statistic for the test of homogeneity. The Pearson Chi-Square was also the test statistic for the test of independence that was discussed earlier. The test statistic is associated with *1* degree of freedom and is found to be *6.529*.

Chi-Square Tests

	Value	df	Asymp. Sig. (2-sided)	Exact Sig. (2-sided)	Exact Sig. (1-sided)
Pearson Chi-Square	6.529[b]	1	.011		
Continuity Correction[a]	6.184	1	.013		
Likelihood Ratio	6.662	1	.010		
Fisher's Exact Test				.011	.006
Linear-by-Linear Association	6.524	1	.011		
N of Valid Cases	1200				

a. Computed only for a 2x2 table

b. 0 cells (.0%) have expected count less than 5. The minimum expected count is 110.67.

Figure 11 - 13

The P-value is *0.011* is less than the significance level. Therefore, we reject the null hypothesis and conclude that the proportion of men who agree when the interviewer is a woman is different from when the interviewer is a man. Although this statistical analysis cannot be used to justify any statement about causality, it does appear that men are influenced by the gender of the interviewer.

Section 11-3 Exercises

1. The number π is an irrational number with the property that when we try to express it in decimal form, it requires an infinite number of decimal places and there is no pattern of repetition. Table 11 - 4 shows how the frequencies of the first *100* digits of π occur. At the *0.05* significance level, test the claim that the digits are uniformly distributed (all have the same relative frequency of *10%*).

Digit	0	1	2	3	4	5	6	7	8	9
Frequency	8	8	12	11	10	8	9	8	12	14

Table 11 - 4

2. Table 11 - 5 gives the observed frequency distribution of a sample of data suspected of coming from a binomial distribution with $n = 3$ and $p = 1/3$.

Number of successes	0	1	2	3
Frequency	89	133	52	56

Table 11 - 5

a. Find the probability corresponding to each category of the table. Hint: See Section 5-2 of this manual.

b. Use a *0.05* significance level to test the claim that the observed frequencies fit a binomial distribution for which $n = 3$ and $p = 1/3$.

3. A survey was conducted to determine whether there is a gender gap in the confidence people have in police. The sample results are listed in Table 11 - 6. Use a *0.05* significance level to test the claim that there is such a gender gap.

	Confidence in Police		
	Great Deal	Some	Very Little or None
Men	115	56	29
Women	175	94	31

Table 11 - 6

4. Is political philosophy (liberal, moderate, conservative) of an individual independent of the affiliated news stations (ABC, CBS, NBC, FOX) the individual watches? A survey of *450* people was conducted to determine if the affiliated news stations they watched were independent of their political philosophy. The data from the survey are shown in Table 11 - 7. Use a *0.05* significance level to test if political philosophy and news station watched are dependent.

	ABC	CBS	NBC	FOX
Liberal	32	33	52	17
Moderate	46	42	68	20
Conservative	37	35	51	17

Table 11 - 7

5. In a greenhouse experiments involving crossing two hybrid African violets the results shown in Table 11 - 8 are observed. Are these results consistent with the experiment proportions 9 : 3 : 3 : 1?

Blue flower, Green stigma	Blue flower, Red stigma	Red flower, Green stigma	Red flower, Red stigma
120	49	36	12

Table 11 - 8

Chapter 12

Analysis of Variance

Chapter 12   Analysis of Variance

In Chapter 9 we discussed inferences about two means. This chapter focuses on inferences about three or more means. **Analysis of variance** (ANOVA) is a procedure that tests for the equality of three or more means. This is a generalization of inferences about two means: independent samples discussed in Section 9-1 of this manual. If differences exist among the means then a multiple comparisons procedure can be used to determine which means differ. Section 12-1 discusses **one-way analysis of variance** and Section 12-2 discusses **two-way analysis of variance**. *Essentials of Statistics*, 3/e does not discuss two-way analysis of variance. You should be familiar with Chapter 12 of *Elementary Statistics*, 10/e or *Essentials of Statistics*, 3/e prior to beginning this chapter.

Section 12-1 One-way ANOVA

Analysis of Variance (ANOVA) provides methods for testing the equality of three or more population means, that is, the means of a single variable from several populations. **One-way analysis of variance** is a procedure used to test if the means of several (say k) independent random samples are equal or not all equal to each other. The null and alternative hypotheses being tested are written as

H_0: $\mu_1 = \mu_2 = \ldots = \mu_k$
H_1: at least one mean is unequal to the other means

A quantitative variable (the **dependent variable**) contains measurements from k independent random samples. A *single* categorical variable that is used to divide the cases of the dependent variable into k different random samples is called the **factor variable** (or **treatment variable**). The term *one-way* refers to the fact that only one factor variable is present. The means of the k groups defined by the factor variable are called **factor means**. The k groups are called the **factor levels** of the factor variable.

Table 12 - 1 listed on the next page gives sample data on injuries to car crash dummies. This data resulted from car crash experiments conducted by the National Transportation Safety Administration. New cars were crashed into a fixed barrier at *35* miles per hour and several different measurements (**head**, **chest**, **femur**) were recorded to determine how severe an injury might result for the test dummy in the driver's seat.

Test the hypothesis that the mean number of Head Injuries for the four different sizes of cars comes from populations with the same mean. That is, test the null and alternative hypotheses that

H_0: $\mu_{\text{Subcompact}} = \mu_{\text{Compact}} = \mu_{\text{Midsize}} = \mu_{\text{Full-size}}$
H_1: at least one mean is unequal to the other means

Enter the data in Table 12 - 1 into a new data file. Create four variables, *size*, *head*, *chest*, and *femur*. The variable *size* has values *1*, *2*, *3*, and *4*, associated with the value labels: Subcompact, Compact, Midsize, and Full-size, respectively. The variable *size* is the categorical variable that separates the independent variables into their respective factor levels. Save the data file to the file **Car Crash Dummies.sav**; the data will be used again in the exercises at the end of this chapter and in Section 13-4 of this manual.

Injuries to Car Crash Dummies

	Head Injury (hic)	Chest Deceleration (g)	Left Femur Load (lb)
Subcompact Cars			
Ford Escort	681	55	595
Honda Civic	428	47	1063
Hyundai Accent	917	59	885
Nissan Sentra	898	49	519
Saturn SL4	420	42	422
Compact Cars			
Chevrolet Cavalier	643	57	1051
Dodge Neon	655	57	1193
Mazda 626 DX	442	46	946
Pontiac Sunfire	514	54	984
Subaru Legacy	525	51	584
Midsize Cars			
Chevrolet Camaro	469	45	629
Dodge Intrepid	727	53	1686
Ford Mustang	525	49	880
Honda Accord	454	51	181
Volvo S70	259	46	645
Full-size Cars			
Audi A8	384	44	1085
Cadillac Deville	656	45	971
Ford Crown Victoria	602	39	996
Oldsmobile Aurora	687	58	804
Pontiac Bonneville	360	44	1376

Table 12 - 1

Test the hypothesis by choosing **Analyze > Compare Means > One-Way ANOVA...** to open the **One-Way ANOVA** dialog box (Figure 12 - 1).

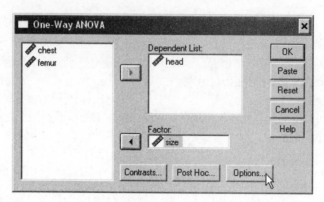

Figure 12 - 1

The variables *head*, *chest*, and *femur* are all dependent variables and should be analyzed separately. This problem is concerned with comparing the mean numbers of Head Injuries for each size of car, thus we will analyze the variable *head*. The categorical variable *size* divides the variable *head* into categories; it is the factor variable. Paste the variable *head* into the **Dependent List** box and the variable *size* into the **Factor** box.

Click the **Options...** button to open the **One-Way ANOVA: Options** (Figure 12 - 2) dialog box. Choose the checkbox for **Descriptive** to display descriptive summary statistics for each group defined by the factor variable *size*. The descriptive statistics displayed are the number of cases, mean, standard deviation, standard error, minimum, maximum, and a *95%* confidence interval for the dependent variable *head* in each group.

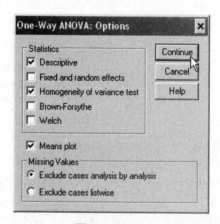

Figure 12 - 2

Choose the checkbox for **Homogeneity of variance test** to check for violations of the equal variance assumption using the Levene's test for equality of variances. This test is not the same that was discussed in your textbook (see Section 9-4 of this manual) to compare the variances two samples. This test compares the variances of several samples.

Choose the checkbox for **Means plot** to obtain a plot of the means for each group defined by the variable *size*. If we determine that differences exist among the means (that is, we reject the null hypothesis) then the means plot helps explain how the means differ from each other.

Click the **Continue** button to close the **One-Way ANOVA: Options** dialog box. Then click the **OK** button to display the results of the ANOVA in the Output Viewer window. The Descriptives table for the variable ***head*** (Figure 12 - 3) shows summary statistics for the data combined and for each factor level defined by the variable *size*. A *95%* confidence interval for each factor mean is also given. The table also indicates that car crash dummies in smaller cars tended to have more head injuries than those in the larger cars, with the possible exception of full-size cars.

HEAD   **Descriptives**

	N	Mean	Std. Deviation	Std. Error	95% Confidence Interval for Mean		Minimum	Maximum
					Lower Bound	Upper Bound		
Subcompact	5	668.80	241.956	108.206	368.37	969.23	420	917
Compact	5	555.80	90.954	40.676	442.87	668.73	442	655
Midsize	5	486.80	167.661	74.980	278.62	694.98	259	727
Full-size	5	537.80	154.613	69.145	345.82	729.78	360	687
Total	20	562.30	172.253	38.517	481.68	642.92	259	917

Figure 12 - 3

The Test of Homogeneity of Variances table (Figure 12 - 4) gives Levene's Statistic for testing the assumption that the variances of the four groups of cars are equal. SPSS reports the P-value of the test statistics to be *0.162*. This is a large P-value, which indicates that the variances of the groups are all equal. Had we rejected this hypothesis then we would not be able to continue with the **ANOVA** procedure. In that case, we would have continued with the **Kruskal-Wallis Test** that will be discussed in Section 13-4 of this manual.

HEAD   **Test of Homogeneity of Variances**

Levene Statistic	df1	df2	Sig.
1.955	3	16	.162

Figure 12 - 4

The ANOVA table (Figure 12 - 5) gives the information needed to test the hypothesis that the factor means are all equal. The Mean Square column gives the **variance between samples** (groups) to be *29475.00* and the **variance within samples** (groups) to be *29707.70*. The F test statistic is based on *3* degrees of freedom in the numerator and *16* degrees of freedom in the denominator and is found to be F = *0.992*. The P-value associated with the F test is *0.422*, therefore we fail to reject the hypothesis and conclude the mean number of head injuries is the same for each class of car.

HEAD   **ANOVA**

	Sum of Squares	df	Mean Square	F	Sig.
Between Groups	88425.000	3	29475.000	.992	.422
Within Groups	475323.2	16	29707.700		
Total	563748.2	19			

Figure 12 - 5

The Means plot (Figure 12 - 6) displays the mean number of head injuries for each factor level (subcompact, compact, midsize, and full-size) of the variable *size*. There are large differences in the mean number of head injuries for each factor level. Nevertheless, beware of making decisions not based on statistical hypotheses. Large looking differences in the factor means may turn out to be non-significant as they did in the case. In this case, the sample sizes were probably too small to discover any significant differences.

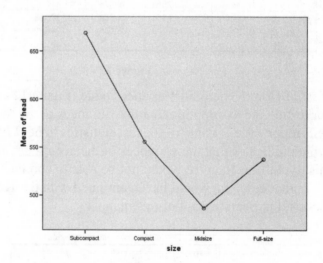

Figure 12 - 6

Section 12-2 Two-way ANOVA*

One-way analysis of variance uses *one* factor to divide the variable into groups. **Two-way analysis of variance** uses *two* factors to divide the variable into subcategories (groups). Table 12 - 2 lists the weights of poplar trees in kg categorized by two variables, *site* (soil conditions) and *treatment*. It is likely that both the treatment and site of a poplar tree will have an effect on the weight of the tree. We will use two-way analysis of variance to determine if there is an interaction effect between treatment and site, if there is an effect for site, and if there is an effect for treatment.

| Poplar Tree Weights (kg) | | | |
No Treatment	Fertilizer	Irrigation	Fertilizer and Irrigation
Site 1			
0.15	1.34	0.23	2.03
.0.02	0.14	0.04	0.27
.016	0.02	0.34	0.92
0.37	0.08	0.16	1.07
0.22	0.08	0.05	2.38
Site 2			
0.06	1.16	0.62	.022
1.11	0.93	0.08	2.13
0.07	0.30	0.62	2.33
0.07	0.59	0.01	1.74
0.44	0.17	0.03	0.12

(Site 1 = rich, moist; Site 2 = sandy, dry)

Table 12 - 2

In a new data file create three variables, *treatment*, *site*, and *weight*. The variable *treatment* should take on the values *1*, *2*, *3*, and *4* associated with the value labels: No Treatment, Fertilizer, Irrigation, and Fertilizer / Irrigation, respectively. The variable *site* should take on the values *1* and *2*, associated with the value labels: Rich/Moist and Sandy/Dry, respectively. The value *weight* should take on the values in Table 12 - 2.

To do a two-way analysis of variance, choose **Analyze > General Linear Model > Univariate...** to open the **Univariate** dialog box (Figure 12 - 7).

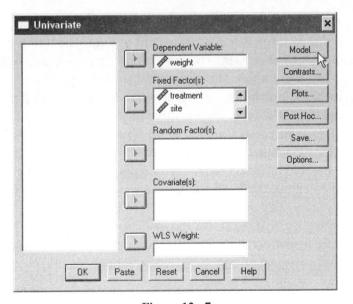

Figure 12 - 7

Paste the variable *weight* into the **Dependent Variable** box. Paste the two variables, *treatment* and *site*, into the **Fixed Factor(s)** box. Click the **Model...** button to open the **Univariate: Model** dialog box (not shown). In this dialog box, uncheck the checkbox for **Include intercept in model**. This step is optional; it makes the ANOVA table displayed by SPSS more similar to the one discussed in *Elementary Statistics*, 10/e. Click the **Continue** button to return to the **Univariate** dialog box.

Click the **OK** button to display the two-way analysis of variance results in the Output Viewer window. The Tests of Between-Subjects Effects table (Figure 12 - 8) shows the two-way analysis of variance table. Compare with the Minitab display shown in Chapter 12-3 of *Elementary Statistics*, 10/e.

Tests of Between-Subjects Effects

Dependent Variable: weight

Source	Type III Sum of Squares	df	Mean Square	F	Sig.
Model	21.727ᵃ	8	2.716	8.102	.000
treatment	7.547	3	2.516	7.505	.001
site	.272	1	.272	.812	.374
treatment * site	.172	3	.057	.171	.915
Error	10.727	32	.335		
Total	32.453	40			

a. R Squared = .669 (Adjusted R Squared = .587)

Figure 12 - 8

There is no interaction between the two factors. The F test statistic is *.171* and the corresponding P-value is *0.915*, so we fail to reject the null hypothesis. We conclude that poplar tree weights are not affected by an interaction between site and treatment.

There is no effect due to site either. The F test statistic is *.812* and the corresponding P-value is *0.374*, so we fail to reject the null hypothesis. We conclude that the site does not appear to have an effect on poplar tree weight. It would seem likely that tree weight would be improved by one type of soil condition over another. Perhaps there are either not enough sample values for the effect to be significant or maybe poplar trees are just not sensitive to soil conditions.

There is an effect due to treatment. The F test statistic is *7.505* and the corresponding P-value is *0.001*, so we reject the null hypothesis. That means we conclude that the treatment received by a poplar tree has an effect on the tree's weight. We can use the **Explore** procedure to obtain the mean weights for each treatment group. The mean weight is *0.321* for the none category, it is *0.481* for the fertilizer category, it is 0.221 for the irrigation category, and it is *1.321* for the fertilizer and irrigation category. We can see from these values that the mean weights are increasing with additional beneficial treatments.

Section 12-3 Exercises

1. Use the Injuries to Car Crash Dummies data in Table 12 - 1 to test the hypothesis that the mean numbers of Chest Deceleration injuries of the four groups (subcompact, compact, midsize, full-size) of cars are equal. Test the hypothesis using a *0.05* significance level.

2. Use the Injuries to Car Crash Dummies data in Table 12 - 1 to test the hypothesis that the mean numbers of Left Femur Load injuries of the four groups (subcompact, compact, midsize, full-size) of cars are equal. Test the hypothesis using a *0.05* significance level.

3. Data Set 13: **Weights of a Sample of M&M Plain Candies** in Appendix B of your textbook (this data is saved on the data disk as **M&M.sav**) lists the weights in grams of red, orange, yellow, brown, blue, and green M&Ms found in a sample of *100* M&Ms. Test the claim that the mean weight of M&Ms is the same for each of the six different color populations using a *0.05* significance level. If it is the intent of Mars, Inc. to make the candies so that the different color populations have the same mean weight, do these results suggest that the company has a problem requiring corrective action?

4. The following table lists pulse rates from Data Set 1 **Health Exam Results** in Appendix B of your textbook (this data is saved on the data disk as **Mhealth.sav**). Are pulse rates affected by an interaction between gender and age? Are pulse rates affected by gender? Are pulse rates affected by age?

	Age		
	Under 20	20 – 40	Over 40
Male	96 64 68 60	64 88 72 64	68 72 60 88
Female	76 64 76 68	72 88 72 68	60 68 72 64

Chapter 13

Nonparametric Statistics

Chapter 13   Nonparametric Statistics

The inferential techniques (e.g., hypothesis testing and confidence intervals) discussed until now have all been parametric procedures. **Parametric tests** require assumptions about the population being sampled. **Nonparametric tests** do not require assumptions about the population distributions. Consequently, nonparametric tests are sometimes called **distribution-free tests**. All of the parametric tests we have discussed have assumed that the data were sampled from a population that follows a normal distribution. In this chapter, we will discuss techniques that can be used when we are unable to assume normality. You should read Chapter 13 of *Elementary Statistics*, 10/e or *Essentials of Statistics*, 3/e prior to beginning this chapter.

Section 13-1   Sign Test

The **Sign test** is a nonparametric (distribution-free) test that tests claims about matched pairs of sample data, nominal data, and the median of a sample. The **Two-Related-Samples Tests** procedure computes the Sign test. This procedure is very similar to the *parametric* **Two-Related Samples T Test** procedure that was discussed in Section 9-3 of this manual.

We will show how to use the Sign test to test a claim about the medians of two paired samples (matched pairs) and how to test a claim about the median of a single sample. In both situations, differences are computed based on the claim and classified as either positive, negative, or tied. These differences (ignoring ties) are used to determine the test statistic (e.g. the smaller of the number of positive difference and the number of negative differences) and the P-value for the test. The exact P-value is determined using the binomial distribution when there are *25* or fewer cases. If there are more than *25* cases, then the P-value is determined using the normal approximation discussed in Section 6-3 of this manual.

Claims involving Matched Pairs

Consider the data listed in Table 9-3 of Section 9-3 of this manual. This data lists the reported height and measured height of *11* male students in a Statistics class. Because the sample size is small, it is not reasonable to assume the population of reported heights and measured heights come from a normal population. Therefore, we will use the Sign test, a nonparametric (distribution-free) test, to decide if that the median of the reported heights is greater than the median of the measured heights.

That is, we will test the hypothesis that

H_0: the median of the reported heights equals the median of the measured heights

H_1: the median difference (reported heights – measured heights) is greater than zero

using a *0.05* significance level.

Open the **Height.sav** data file (this data file was saved in Section 9-3 of this manual). If the **Height.sav** data file is not available, enter the *11* pairs of data values in Table 9-3, into the two variables *reported* and *measured*.

Choose **Analyze > Nonparametric Tests > 2 Related Samples…** to open the **Two-Related-Samples Tests** dialog box (Figure 13 - 1).

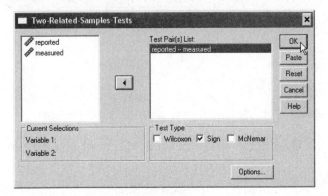

Figure 13 - 1

First, click on *reported* and then on *measured*. This will select *reported* as Variable 1 and *measured* as Variable 2 in a similar manner as was used with the **Paired-Samples T Test** procedure in Section 9-3. Once the variables have been selected, click the **variable paste** button and SPSS will create the difference *reported – measured* in the **Test Pair(s) List** box. Choose the checkbox for **Sign** and click the **OK** button. The results of the **Sign test** will appear in the Output Viewer window. The Frequencies table (Figure 13 - 2) reports that there are *10* negative differences, *1* positive difference, and no ties.

Frequencies

		N
MEASURED - REPORTED	Negative Differences[a]	10
	Positive Differences[b]	1
	Ties[c]	0
	Total	11

a. MEASURED < REPORTED

b. MEASURED > REPORTED

c. REPORTED = MEASURED

Figure 13 - 2

The Test Statistics table (Figure 13 - 3) reports the P-value for the Sign test to be *0.012*. Recall that SPSS always calculates the P-value for the two-tailed alternative. Since the claim being tested is one-tailed alternative, we need to divide this P-value by two. The P-value of the hypothesis test is therefore *0.006*. The P-value was computed exactly using the binomial distribution because the number of cases was less than *25*. The P-value is very small; therefore, we reject the null hypothesis and conclude that the median difference between the reported heights and the measured heights is greater than zero. The male Statistics students are likely exaggerating their heights.

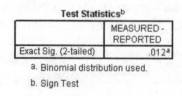

Test Statistics[b]

	MEASURED - REPORTED
Exact Sig. (2-tailed)	.012[a]

a. Binomial distribution used.

b. Sign Test

Figure 13 - 3

Claims about the Median of a Single Population

Consider the Data Set 2: **Body Temperatures of Healthy Adults** listed in Appendix B of your textbook (this data is saved on the data disk as **BODYTEMP.sav**). This data lists the body temperatures in degrees Fahrenheit of a random sample of *105* healthy adults. Use the sample data to test the claim, using a *0.05* significance level, that the median body temperature is equal to *98.6°F*. That is, we will test the hypothesis that

H_0: the median body temperature of an adult is *98.6°F*

H_1: the median body temperature of an adult is not equal to *98.6°F*

using a *0.05* significance level.

Open the **BODYTEMP.sav** data file. There is only one variable in the data file. In order, to do this test we must create another variable *constant* that has the value of *98.6* for each case. The easiest way to do this is to choose **Transform > Compute** to open the **Compute Variable** dialog box. Type *constant* into the **Target Variable** box and *98.6* into the **Numeric Expression** box, then click the **OK** button. The data file will now include a new variable *constant* that is equal to *98.6* for every case.

To do the hypothesis test, choose **Analyze > Nonparametric Tests > 2 Related Samples...** to open the **Two-Related-Samples Tests** dialog box (Figure 13 - 1). First, click on *BodyTemp* and then on *constant*. This will select *BodyTemp* as Variable 1 and *constant* as Variable 2. Once the variables have been selected, click the **variable paste** [▶] button and SPSS will create the difference *weight – constant* in the **Test Pair(s) List** box. Choose the checkbox for **Sign** and click the **OK** button. The results of the **Sign test** will appear in the Output Viewer window.

The Frequencies table (Figure 13 - 4) reports that there are *23* negative differences, *68* positive differences, and *14* ties.

Frequencies

		N
constant - BodyTemp	Negative Differences[a]	23
	Positive Differences[b]	68
	Ties[c]	14
	Total	105

a. constant < BodyTemp

b. constant > BodyTemp

c. constant = BodyTemp

Figure 13 - 4

The Test Statistics table (Figure 13 - 5) reports the P-value for the Sign test to be *0.000*. The P-value was computed using the normal approximation since there were more than *25* cases. The P-value is very small; therefore, we reject the null hypothesis and conclude that the median body temperature for a healthy adult is not *98.6°F*.

Test Statistics[a]

	constant - BodyTemp
Z	-4.612
Asymp. Sig. (2-tailed)	.000

a. Sign Test

Figure 13 - 5

Section 13-2 Wilcoxon Signed-Ranks Test for Matched Pairs

The **Wilcoxon Signed-Ranks test** is a nonparametric (distribution-free) procedure used with two related variables to test the hypothesis that the two variables have the same distribution. This test takes into account information about the magnitude of differences within pairs and gives more weight to pairs that show large differences than to pairs that show small differences. The test statistic is based on the ranks of the absolute values of the differences between the two variables. The Wilcoxon Signed-Ranks test can be used in the same situations as the Sign test. It usually yields better conclusions than the Sign test because it uses more information than the Sign test.

In this section, we repeat the analysis of the heights of the male Statistics students. This time we use the Wilcoxon Signed-Ranks test to test the hypothesis that the reported heights are greater than the measured heights. Because the **Wilcoxon Signed-Ranks** procedure uses more information in its calculations, claim being tested in now substantially stronger. Namely, we are now testing the null and alternative hypotheses that:

H_0: The two samples come from populations with the same distribution

H_1: The two samples come from populations with different distributions

using a *0.05* significance level. To test this hypothesis, follow the exactly same procedure as in Section 13-1 above except choose the bullet for **Wilcoxon** in Figure 13 - 1 instead of the bullet for **Sign**.

The Ranks table (Figure 13 - 6) now includes information about the Mean Rank and Sum of Ranks, which are useful in calculating the Wilcoxon Signed-Ranks test statistic. The sum of the absolute values of the negative ranks is *60* and the sum of the absolute value of the positive ranks is *6*. The Wilcoxon Signed-Ranks test statistics is the smaller of these two numbers, that is $T = 6$.

Ranks

		N	Mean Rank	Sum of Ranks
MEASURED - REPORTED	Negative Ranks	10[a]	6.00	60.00
	Positive Ranks	1[b]	6.00	6.00
	Ties	0[c]		
	Total	11		

a. MEASURED < REPORTED

b. MEASURED > REPORTED

c. REPORTED = MEASURED

Figure 13 - 6

The Test Statistics table (Figure 13 - 7) indicates that Normal approximation test statistic is *-2.408* and the corresponding P-value is *0.016*. Therefore, we reject the null hypothesis and conclude that the two samples come from populations with different distributions.

Test Statistics[b]

	MEASURED - REPORTED
Z	-2.408[a]
Asymp. Sig. (2-tailed)	.016

a. Based on positive ranks.

b. Wilcoxon Signed Ranks Test

Figure 13 - 7

Section 13-3 Wilcoxon Rank-Sum Test for Two Independent Samples

The **Two-Independent Samples Tests** procedure provides four nonparametric statistical tests, the **Mann-Whitney test** (also known as the **Wilcoxon Rank-Sum test**), the **Moses extreme reactions test**, the **Kolmogorov-Smirnov Z test**, and the **Wald-Wolfowitz runs test**. The Wilcoxon Rank-Sum test is a nonparametric test that is equivalent to the independent-samples *t*-test discussed in Section 9-2 of this manual. It tests whether two independent samples are from the same population. The other three tests are more general than the Mann-Whitney test and are not discussed in your textbook.

Consider Data Set 1: **Health Exam Results** listed in Appendix B of your textbook (this data is saved on the data disk as **Mhealth.sav** - men and **Fhealth.sav** - women). Select only the first 13 sample values of the body mass index (BMI) of men and the first 12 sample values of BMI of women and copy to a new data file (refer to Section 1 – 2 for the procedure to select a range of cases, or simply use copy and paste). Create a new variable called *gender* and assign labels of *1 = Male* and *2 = Female* for the male and female sample values. Use a 0.05 significance level to test the claim that the median BMI of men is equal to the median BMI of women. The claim requires that we test the null and alternative hypotheses that

H_0: Men and women have BMI values with equal medians.

H_1: Men and women have BMI values with medians that are not equal.

Choose **Analyze > Nonparametric Tests > 2 Independent Samples...** to open the **Two-Independent-Samples Tests** dialog box (Figure 13 - 8) will open. Paste the variable *bmi* into the **Test Variable List** box and the variable *gender* into the **Grouping Variable** box.

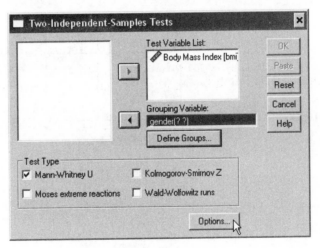

Figure 13 - 8

Next, click the **Define Groups...** button to open the **Two Independent Samples: Define Groups** dialog box (Figure 13 - 9). Enter *1* (value associated with the value label for male) into the **Group1** box and 2 (value associated with value label for female) into the **Group 2** box.

Figure 13 - 9

Choose the checkbox for **Mann-Whitney** U and click the **OK** button. The results will appear in the Output Viewer window. The Ranks table (Figure 13 - 10) gives the sum of the ranks for the 13 male BMI values as *187.00* and the sum of the ranks for the 12 female BMIs as *138.00*.

Ranks

	gender	N	Mean Rank	Sum of Ranks
Body Mass Index	Male	13	14.38	187.00
	Female	12	11.50	138.00
	Total	25		

Figure 13 - 10

The Test Statistics table (Figure 13 - 11) reports both the Mann-Whitney U test statistic ($U = 60.000$) and the Wilcoxon Rank-Sum test statistic ($W = 138.00$). The normal approximation test statistic is *-0.980* and has a corresponding P-value of *0.327*. The exact P-value is *0.347*. Therefore, we fail to reject the null hypothesis and conclude that the distributions of the BMS values of males and females are basically the same.

Test Statistics[b]

	Body Mass Index
Mann-Whitney U	60.000
Wilcoxon W	138.000
Z	-.980
Asymp. Sig. (2-tailed)	.327
Exact Sig. [2*(1-tailed Sig.)]	.347[a]

a. Not corrected for ties.

b. Grouping Variable: gender

Figure 13 - 11

Section 13-4 Kruskal-Wallis Test

The Kruskal-Wallis test is a nonparametric alternative to the one-way ANOVA discussed in Section 12-1 of this manual. The **Kruskal-Wallis** procedure tests whether several independent samples are from the same population. The Kruskal-Wallis test applies when the distributions (one for each population) of the variable under consideration have the same shape, but does not require the distributions to be normal or have any other specific shape. The Kruskal-Wallis test is discussed in Chapter 13-5 of your textbook.

Table 12 - 1 lists sample data on Injuries to Car Crash Dummies. This data resulted from car crash experiments conducted by the National Transportation Safety Administration. New cars were crashed into a fixed barrier at *35* miles per hour and several different measurements (***head***, ***chest***, ***femur***) were recorded to determine how severe an injury might result for the test dummy in the driver's seat. Using the head injury measurements, is there sufficient evidence to conclude that head injuries for the four weight categories of cars are not all the same? That is, test the claim that

H_0: $\mu_{Subcompact} = \mu_{Compact} = \mu_{Midsize} = \mu_{Full\text{-}size}$
H_1: At least one of the means is unequal to the other means

Open the **Car Crash Dummies.sav** data file (this data file was saved in Section 12-1 of this manual). If the **Car Crash Dummies.sav** data file is not available, then enter the data for head injury (*head*) and weight class (*size*) from Table 12 - 1 into SPSS. The variable *size* has values *1*, *2*, *3*, and *4*, associated with the value labels: Subcompact, Compact, Midsize, and Full-size, respectively.

Choose **Analyze > Nonparametric Tests > K Independent Samples…** to open the **Tests for Several Independent Samples** dialog box (Figure 13 - 12). Paste the variable *head* into the **Test Variable List** and the variable *size* into the **Grouping Variable** box.

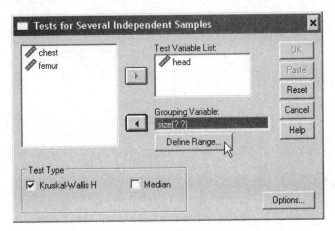

Figure 13 - 12

Click the **Define Range…** button to open the **Several Independent Samples: Define Range** dialog box (Figure 13 - 13). Define the range by entering the minimum and maximum values that correspond to the lowest and highest categories of the grouping variable. In our case, enter *1* and *4*. Cases with values outside of these bounds will be excluded from the analysis.

Figure 13 - 13

Click the **Continue** button. Choose the checkbox for **Kruskal-Wallis H** and click the **OK** button. The results will be displayed in the Output Viewer window.

The Ranks (Figure 13 - 14) table shows the Mean Rank for each weight class. We can see that generally, the mean rank is larger for smaller cars, that is, the number of head injuries is generally greater for smaller cars than larger cars.

Ranks

	SIZE	N	Mean Rank
HEAD	Subcompact	5	12.80
	Compact	5	10.50
	Midsize	5	8.90
	Full-size	5	9.80
	Total	20	

Figure 13 - 14

The Test Statistics table (Figure 13 - 15) gives the Kruskal-Walls H test statistic (Chi-Square) as *1.192* with associated degrees of freedom of *3*. The P-value of this test statistic is given as *0.755*; therefore, we fail to reject the null hypothesis and conclude that the mean number of head injuries within each of the weight classes are equal.

Test Statistics[a,b]

	HEAD
Chi-Square	1.192
df	3
Asymp. Sig.	.755

a. Kruskal Wallis Test

b. Grouping Variable: SIZE

Figure 13 - 15

Section 13-5 Rank Correlation

In Chapter 10, we saw that correlation measures the strength of the *linear* relationship between two variables. **Spearman's rank correlation coefficient, r_s** is a nonparametric version of **Pearson's correlation coefficient, r** discussed in Section 10-1 of this manual. Unlike Pearson's correlation coefficient, there is no requirement that the sample pairs of data have a bivariate normal distribution. It is based on the ranks of the data rather than the actual values and is appropriate for ordinal data.

As with the Pearson correlation coefficient, the sign of the coefficient indicates the direction of the relationship, and the larger the absolute value the stronger the relationship. The two variables are called **negatively linearly correlated** when r_s is negative, **positively linearly correlated** when r_s is positive, and **uncorrelated** when r_s is zero. Correlation coefficients that are close to *–1* or *+1* indicate a strong linear relationship between the two variables.

In Section 10-1 of this manual, we discussed the problem of determining how much of a gratuity to leave a waitress or waiter in a restaurant. We found the Pearson's correlation coefficient between the amount of the bill and the amount of the gratuity for a sample of six meals. The data is listed in Table 10 - 1 of this manual. Because the sample size is small, it is not reasonable to assume the sample pairs of data have a bivariate normal distribution. Therefore, we will determine the correlation using Spearman's rank correlation coefficient and test the claim that

$$H_0: \rho_s = 0$$
$$H_1: \rho_s \neq 0$$

using a *0.05* significance level. In a new data file create two variables, **bill** and **tip**, that have data type *Dollar7.2*. Enter the six pairs of data values in Table 10 - 1 into the file.

The **Bivariate Correlations** dialog box that we used in Section 10-1 to determine Pearson's correlation coefficient also calculates Spearman's rank correlation coefficient. To determine Spearman's rank correlation coefficient, choose **Analyze > Correlate > Bivariate...** to open the **Bivariate Correlations** dialog box (Figure 10 - 1). Paste the two variables **bill** and **tip** into the variables box. Choose the checkbox for **Spearman** and then click the **OK** button. The results will appear in the Output Viewer window.

The Correlations table (Figure 13 - 16) reports Spearman's rank correlation coefficient between the amount of the restaurant bill and the amount of the gratuity to be *0.829*. This value is very close to the Pearson's correlation coefficient *0.828* that we determined in Section 10-1.

Correlations

			Amount of restaurant bill	Amount of Gratuity
Spearman's rho	Amount of restaurant bill	Correlation Coefficient	1.000	.829*
		Sig. (2-tailed)	.	.042
		N	6	6
	Amount of Gratuity	Correlation Coefficient	.829*	1.000
		Sig. (2-tailed)	.042	.
		N	6	6

*. Correlation is significant at the .05 level (2-tailed).

Figure 13 - 16

SPSS also displays the P-value associated with testing the hypothesis that Spearman's rank correlation coefficient is different from zero to be *0.042*. Therefore, we reject the null hypothesis and conclude that the Spearman's rank correlation coefficient is different from zero.

Section 13-6 Runs Test for Randomness

The **Runs test** is used to determine if the order of occurrence of sample data values is random. A **run** is a sequence of like observations. There is no parametric equivalent to this test.

In the course of a WNBA game, Cynthia Cooper shoots *12* free throws. Denoting shots made by H (for "hit") and denoting shots missed by M, her results are as follows: H, H, H, M, H, H, H, H, M, M, M, and H. Use a *0.05* significance level to test for randomness in the sequence of hits and misses. The data values in SPSS must be numeric therefore use a *1* for a made shot (a hit) and a *0* for a missed shot. In a new data file, create a new variable *shots* with value labels H and M, associated with *1* and *0*, respectively. Enter the data into the variable *shots*.

Choose **Analyze > Nonparametric Tests > Runs...** to open the **Runs Test** (Figure 13 - 17) dialog box. Paste the variable *shots* into the **Test Variable List** box.

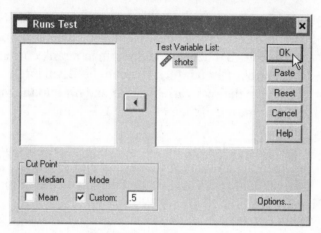

Figure 13 - 17

Choose one or more of the checkboxes in the **Custom** portion of the dialog box to define a cut point. SPSS uses a cut point to dichotomize the variables (all the values less than the cut point have the same characteristic). You can use the observed mean, the median, the mode, or a specified value as a cut point. For this data, any value between *0* and *1* will work, since we want all the *0*'s (misses) in one group and all the *1*'s (hits) in the other group. Type *0.5* into the **Custom** box.

Click the **OK** button and the results will appear in the Output Viewer window. The Runs Test table (Figure 13 - 18) reports the Number of Runs to be *5*. The normal approximation test statistic is *-0.576* with a corresponding P-value of *0.565*. Therefore, we fail to reject the null hypothesis and conclude that the data is random.

Runs Test

	SHOTS
Test Value[a]	.5
Total Cases	12
Number of Runs	5
Z	-.575
Asymp. Sig. (2-tailed)	.565

a. User-specified.

Figure 13 - 18

Section 13-7 Exercises

Exercises 1-2 use the data listed in Data Set 16: **Weights of Discarded Household Garbage for One Week** in Appendix B of your textbook (this data is saved on the data disk as **Garbage.sav**). The data set contains the weights in pounds of discarded metal, paper, plastic, glass, food, yard, textile, and other goods for a sample of households for one week.

1. Use the Sign test to test the claim that the mean weight of discarded metal and mean weight of discarded paper are equal using a *0.05* significance level.

2. Use the **Wilcoxon Signed-Ranks** procedure to test the claim that the mean weight of discarded metal and mean weight of discarded paper are equal using a *0.05* significance level.

3. Data Set 13: **Weights of a Sample of M&M Plain Candies** in Appendix B of your textbook (this data is saved on the data disk as **M&M.sav**) lists the weights of red, orange, yellow, brown, blue, and green M&Ms found in a sample of *100* M&Ms. Use the Wilcoxon Rank-Sum procedure to test the hypothesis that orange and brown M&Ms weigh the same. Use a significance level of *0.05*.

4. Use the **Kruskal-Wallis** procedure to test the hypothesis that the mean numbers of Chest Deceleration injuries of the four groups (subcompact, compact, midsize, full-size) of cars are equal. Use the Injuries to Car Crash Test Dummies data in Table 11 - 1 from Section 12-1 of this manual. Test the hypothesis using a *0.05* significance level.

5. Data Set 6: **Bears (wild bears anesthetized)** in Appendix B of your textbook (this data is saved on the data disk as **Bears.sav**) contains nine measurements on *54* anesthetized wild bears. The nine variables in the data file are: *age* (age in months), *month* (month of measurement), *sex*, *headlen* (length of head in inches), *headwth* (width of head in inches), *neck* (distance around neck in inches), *length* (length of body in inches), *chest* (distance around the chest in inches), and *weight* (measured weight in pounds). Find Spearman's rank correlation between *weight* and the variables *age, headlen, headwth, neck, length,* and *chest*. Which variable is most correlated with *weight*? Which variable is least correlated with *weight*?

6. Trends in business and economics applications are often analyzed with the runs test. The annual high points of the Dow Jones Industrial Average are listed in Table 13 - 1 (in order by row) for a recent sequence of years. Test for randomness above and below the median. Hint: choose the checkbox for the **Median** cut point. What does the result suggest about the stock market as an investment consideration?

969	842	951	1036	1052	892	882	1015	1000	908
898	1000	1024	1071	1287	1287	1553	1956	2722	2184
2791	3000	3169	3413	3794	3978	5216	6561	8259	9338

Table 13 - 1

Chapter 14

Statistical Process Control

Chapter 14   Statistical Process Control

Various statistics used to describe the distribution of a sample were discussed in Chapter 2. When data are collected over time, **control charts** are used to check if the mean or the spread of the data changes over time. Data that is arranged according to some time sequence is called **process data**. This chapter will show how to make control charts in SPSS. You should read Chapter 14 of *Elementary Statistics*, 10/e or *Essentials of Statistics*, 3/e prior to beginning this chapter.

 Control charts are used to monitor some characteristic or attribute of data that might change over time. A control chart allows us to check if the center of the distribution or the spread of the distribution is changing over time. Control charts for quantitative data monitor the mean, range, or standard deviation of a process over time. There are several different kinds of controls charts. In this section, we will discuss the run chart, the R chart, and the $\bar{x}$ chart. A **run chart** is a sequential plot of individual data values over time. An **R chart** (or **range chart**) is a plot of the sample ranges over time; it is used to monitor the variation in a process. An $\bar{x}$ **chart** (or **mean chart**) is a plot of the sample mean over time; it is used to monitor the center in a process.

Section 14-1 Control Charts for Variation and Mean

Run charts

Data Set 15: **Axial Loads of Aluminum Cans** in Appendix B of your textbook (this data is saved on the data disk as **Cans.sav**) lists the axial loads in pounds of aluminum cans. An axial load of an aluminum can is the maximum weight supported by its sides. It is important to have an axial load high enough so that the can isn't crushed when the top lid is pressed into place. During each day of production, seven cans were randomly selected and tested day over a 25-day period. The data set lists the axial loads of two samples of 175 aluminum cans each, one with 0.0109-inch thickness, and the other with 0.0111-inch thickness. Construct a run chart for the 0.0109-inch thickness cans and determine whether the process is within statistical control. If not, identify which of the three out-of-control criteria apply.

 Open the data file **Cans.sav**. You need to create a new variable *day*, which is the day on which the observation was recorded. The first seven observations were recorded on day *1*, the next seven on day *2*, and so on for *25* days. There is no easy way to do this; you will need to type the *175* data values (seven *1*'s, followed by seven *2*'s, followed by seven *3*'s, etc.) into the variable *day*. This defines the day for the *0.0109*-inch thickness cans. You then need to repeat this process for the next *175* data values (the ones associated with *0.0111*-inch thickness cans). It is easiest to highlight the first *175* data values and then copy them into the second *175* values of the variable *day*.

Next, select only the *0.0109*-inch thickness cans by choosing **Data > Select Cases…** to open the **Select Cases** dialog box (Figure 1 - 2). Choose the bullet for **If condition is satisfied** and click the **If…** button to open the **Select Cases: If** dialog box (Figure 1 - 6). Type the expression *sample = 1* into the box and click the **Continue** button. The value labels of a variable can be determined in the Variable View (see Section 0-3 for details). For example, the variable **sample** takes on values *1* and *2*, associated with value labels: 0.0109 in and 0.0111 in, respectively, which the sample.

To create the run chart, choose **Graphs > Control…** to open the **Control Charts** dialog box (Figure 14 - 1). There are four different possible control charts available. The first two options are for quantitative data; the third and fourth options are for qualitative data and will be discussed in the next section. The **X-Bar, R, s** option creates the mean charts and range charts. The **Individuals, Moving Range** option plots individual data values rather than subgroup averages. Choose the icon for **Individuals, Moving Range**.

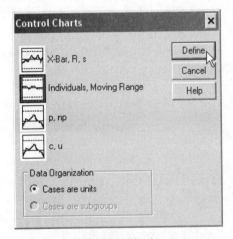

Figure 14 - 1

Click the **Define** button to open the **Individuals and Moving Range** (Figure 14 - 2) dialog box. Paste the variable *load*, labeled *Axial Load*, into the **Process Measurement** box and the variable *day* into the **Subgroups Labeled by** box.

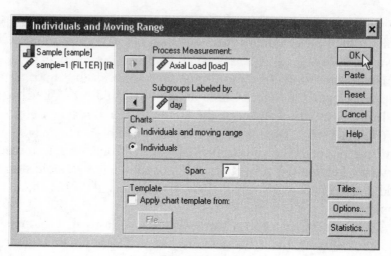

Figure 14 - 2

Choose the bullet for **Individuals** to make a run chart for individuals. Choosing the bullet for **Individuals and moving range** will make a run chart and a moving average chart. Your textbook does not discuss the moving average chart.

The value for **Span** specifies the number of cases used in calculating the control limits for the run chart. Since 7 cans were randomly selected and tested each day, enter the value of 7 into the **Span** box.

Click the **OK** button and the resulting **Run chart** (Figure 14 - 3) will appear in the Output Viewer window.

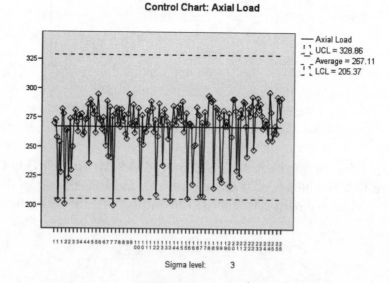

Figure 14 - 3

The process does not appear to be within statistical control. Although there is no obvious trend or shift in the chart, there are at least four data points below the LCL (lower control limit). SPSS automatically calculates the UCL (upper control limit) and LCL by adding and subtracting *3* standard deviations from the average, respectively. We know that observations that are more than three standard deviations from the mean are very unlikely. This process has more values outside the limits than would be expected based on randomness, leading us to believe the process is out of control.

For some processes, it might be more informative to calculate the LCL and UCL based on a different number of standard deviations from the centerline. The number of standard deviations can be specified by clicking the **Options...** button in Figure 14 - 2.

The R chart and $\bar{x}$ chart

Besides checking if the individual values are in statistical control, it is often important to monitor the mean and variation of the process. The $\bar{x}$ **chart** (sometimes called a **mean chart**) is a plot of the sample means and is used to monitor the center of a process. The **R chart** (or **range chart**) is a plot of the sample ranges and used to monitor the variation of the process. SPSS provides an option to plot the sample standard deviation instead of the sample range when making an R chart.

Construct an $\bar{x}$ **chart** and an **R chart** for the *0.0109*-inch thickness cans listed in Data Set 15: **Axial Loads of Aluminum Cans**. Determine whether the process is within statistical control. If not, identify which of the three out-of-control criteria apply.

Open the **Cans.sav** data file. Choose **Graphs > Control...** to open the **Control Charts** dialog box (Figure 14 - 1). If you have not created the variable *day* then create it now (see how to do this in the Run chart section above). Use the **Select Cases** dialog box to select only the *0.0109*-inch thickness cans.

To make an R chart, choose the icon for **X-Bar, R, s**. There are now two choices for **Data Organization**. The **Cases are units** bullet should be chosen when each unit is a separate case in the data file. The number of units can vary between subgroups. The **Cases are subgroups** bullet should be chosen when all units within a subgroup or time interval are recorded in the same case. The number of samples per subgroup should be the same.

In this case, the variable *day* defines the subgroup or time interval to which the unit belongs and the variable *load* records the value being measured. Therefore, choose the bullet for **Cases are units** and click the **Define** button to open the **X-Bar, R, s: Cases Are Units** (Figure 14 - 4) dialog box.

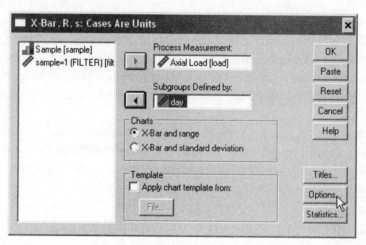

Figure 14 - 4

Paste the variable *load* into the **Process Measurement** box and the variable *day* into the **Subgroups Defined by** box. Choose the bullet for **X-Bar and range** to make a mean chart and range chart. If you prefer to use the standard deviation instead of the range for the R chart, click the bullet for **X-Bar and standard deviation**.

Click the **OK** button and the results will appear in the Output Viewer window. The **Mean chart** (Figure 14 - 5) shows that the mean of the process is within statistical control for *3* standard deviations because (1) there is no obvious pattern, trend, or cycle, (2) there are no points lying outside the control limits and (3) there are not eight consecutive points above or below the centerline.

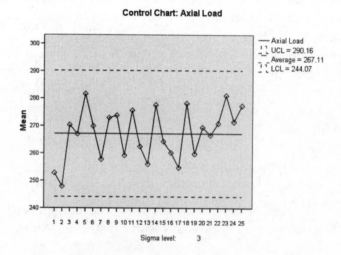

Figure 14 - 5

The **Range chart** (Figure 14 - 6) shows the variation of the process is within statistical control as well. We can conclude this because (1) there is no obvious pattern, trend, or cycle, (2) there are no points outside the upper and lower control limits, and (3) there are not eight consecutive points all above or all below the centerline.

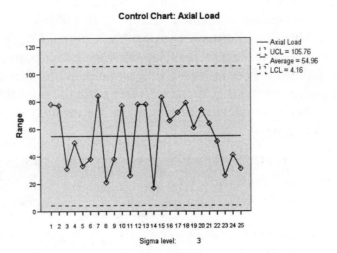

Figure 14 - 6

Section 14-2 Control Charts for Attributes

In this section, we discuss constructing control charts to monitor an attribute (or quality) of sample data. In the previous section, the control charts monitored quantitative data. Control charts for qualitative data monitor the proportion or number of nonconforming items in a process over time. A **nonconforming item** is one that does not meet the specifications or requirements of the process. Nonconforming items are sometimes discarded, repaired, or called *seconds* and sold at reduced prices. A control chart monitoring an attribute is called a **control chart for p** or (**p chart**).

The Altigauge Manufacturing Company produces aircraft altimeters, which provide pilots with readings of their heights above sea level. The accuracy of altimeters is very important to aviation safety because pilots rely on them to maintain altitudes with safe vertical clearances above obstacles. Altimeters are considered to be defective if it cannot be calibrated or corrected to give accurate readings within 20 feet of the true altitude. The Altigauge Manufacturing Company produces altimeters in batches of 100, and each altimeter is tested and determined to be acceptable or defective. Listed in Table 14 -1 are the numbers of defective altimeters in successive batches of 100 (the data also appear in Chapter 14-3 of your textbook). Construct a

control chart for p, the proportion of defective altimeters and determine whether the process is within statistical control. If not, identify which of the three out-of-control criteria apply.

Number of defects: 2   0   1   3   1   2   2   4   3   5   3   7

Table 14 - 1

In a new data file, create a variables, *count..* Enter the *12* numbers in Table 14 - 1 into the variable *count*. Save the data file as **Defective Altimeters.sav**.

Choose **Graphs > Control...** to open the **Control Charts** dialog box (Figure 14 - 1). To make a **p chart**, choose the icon for **p, np**. Choose the bullet for **Cases are subgroups** because each case in our data file corresponds to a batch (a subgroup or time interval). Click the **Define** button to open the **p, np: Cases are Subgroups** dialog box (Figure 14 - 7). Paste the variable *count* into the **Number Nonconforming** box.

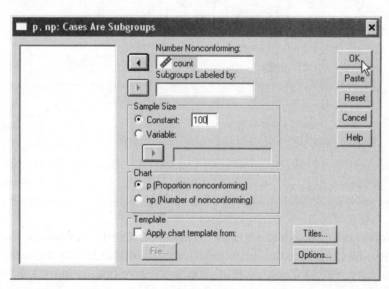

Figure 14 - 7

When the population size is constant, choose the bullet for **Constant** and type the population size into the **Constant** box. Type the value 100 into the **Constant** box. Unfortunately, this box only accepts population sizes less than *99,999*. If our population is larger than this, we would need to create the variable *total* containing the population size for each of the cases and then choose the bullet for **Variable**.

Choose the bullet for **p (Proportion nonconforming)** when the number of samples within each subgroup or time interval is unequal. When the number of samples within a subgroup is equal, you may choose either bullet: **np (Number of nonconforming)** or **p (Proportion nonconforming).** For the p chart below we choose the bullet for **p (Proportion nonconforming)**.

Click the **OK** button and the p chart (Figure 14 - 8) will appear in the Output Viewer window. Although there are no points outside the control limits, the process is out of statistical control because there appears to be an upward trend in the number of defective altimeters. The company should take immediate action to correct the increasing proportion of defects.

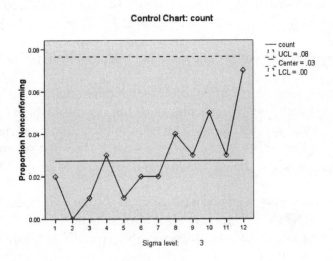

Figure 14 - 8

Section 14-3 Exercises

Exercises 1-3 use the data listed in Data Set 10: **Rainfall (in inches) in Boston for One Year** in Appendix B of your textbook (this data is saved on the data disk as **Bostrain.sav**). The data set contains *365* daily rainfall measurements for Boston.

1. Construct a run chart for the *52* rainfall amounts for Monday. Determine whether the process is within statistical control. If not, identify which of the three out-of-control criteria apply. Hint: Create a filter variable using the **Select Cases** dialog box.

2. Create a variable *week* that indicates the week number. Use the *52* weekly samples of seven values each to construct a mean chart. Determine whether the process is within statistical control. If not, identify which of the three out-of-control criteria apply. Hint: There are *53* cases for Wednesday; therefore, you must delete the *53*[rd] case on Wednesday.

3. Create a variable *week* that indicates the week number. Use the *52* weekly samples of seven values each to construct a range chart. Determine whether the process is within statistical control. If not, identify which of the three out-of-control criteria apply. Hint: There are *53* cases for Wednesday; therefore, you must delete the *53*[rd] case on Wednesday.

Exercises 4 and 5 use the data listed in Data Set 15: **Axial Loads of Aluminum Cans** in Appendix B of your textbook (this data is saved on the data disk as **Cans.sav**) lists the axial loads in pounds of aluminum cans. An axial load of an aluminum can is the maximum weight supported by its sides and it is important to have an axial load high enough so that the can isn't crushed when the top lid is pressed into place. During each day of production, seven cans were randomly selected and tested day over a *25-day* period. The data set lists the axial loads of two samples of *175* aluminum cans each, one with *0.0109*-inch thickness, and the other with *0.0111*-inch thickness.

4. Construct a run chart for the *0.0111*-inch thickness cans. Determine whether the process is within statistical control. If not, identify which of the three out-of-control criteria apply. Hint: You need to create a new variable *day*, which is the day on which the observation was recorded. The first seven observations were recorded on day *1*, the next seven on day *2*, and so on for *25* days.

5. Construct a mean chart and a range chart for the *0.0111*-inch thickness cans. Determine whether the process is within statistical control. If not, identify which of the three out-of-control criteria apply. Hint: You need to create a new variable *day*, which is the day on which the observation was recorded. The first seven observations were recorded on day *1*, the next seven on day *2*, and so on for *25* days.

6. In each of *13* consecutive and recent years, *100,000* children aged 0-4 years were randomly selected and the number who died from infectious diseases is recorded, with the results given below (based on data from "Trends in Infectious Diseases Mortality in the United States," by Pinner et. al. *Journal of the American Medical Association*, Vol. 275, No. 3). Do the results suggest a problem that should be corrected?). Construct a control chart for p and determine whether the process is within statistical control. If not, identify which of the three out-of-control criteria apply.

    Number of deaths: 30   29   29   27   23   25   25   23   24   25   25   24   23

    Table 14 - 2

7. In a continuing study of voter turnout, a sample of *1,000* people of voting age are randomly selected in each of year when there is a national election, and the numbers who actually voted are listed in Table 14 - 3 (based on data from the *Time Almanac*). Construct a control chart for p and determine whether the process is within statistical control. If not, identify which of the three out-of-

control criteria apply.

608   466   552   382   536   372   526   398   531   364   501   365   551   388   491

Table 14 - 3

Appendix A: Answers to Problems

In Appendix A, you will find the answers to all the problems in this manual. The answers are grouped by Chapters according to the problem numbers. This should make it easy for you to find the problems. Most of the time, the answers will involve SPSS output, which is shown. When the problem is solved using the **Compute Variable** dialog box, the expression typed into the **Numeric Expression** window is given as part of the answer.

Chapter 0

1. The Data Editor should look like this when complete:

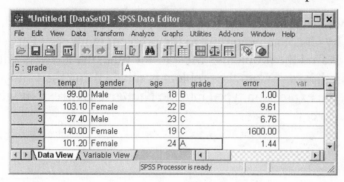

2. a. There are *5* variables– subject, gender, age, distance, and index.
   b. There are *108* cases.
   c. Numeric expression– *12*age*

3. a. There are *7* cases.
   b. *String21*
   c. *Numeric4.0*, just because year is a date does not mean we use the format *Date*.

4. a. There are *2* variables – month and Fahrenheit.
   b. There are *6* cases.
   c. The Data Editor should look like this when complete.

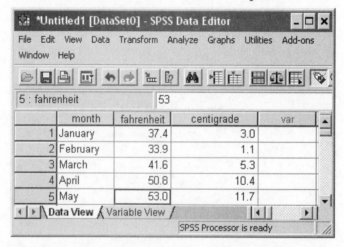

Chapter 1

1. a. ***pathsize*** has value label *Pathologic Tumor Size (cm)*.
      er has value label *Estrogen Receptor Status*.
   b. ***age*** and ***lnpos*** both have Scale measurement level.
      Pathological Tumor Size (Categories) is the value label for the variable
      pathscat that has Ordinal measurement level.
   c. ***histgrad*** with value label *Histologic Grade* should be Nominal.

2. Answers vary because this is a simple random sample.

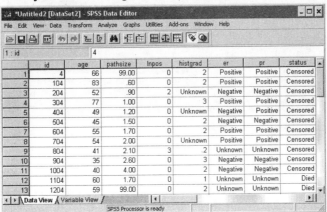

3. The systematic sample is shown below.

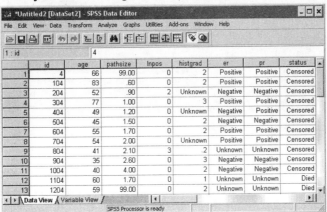

4.   The systematic sample is shown below.

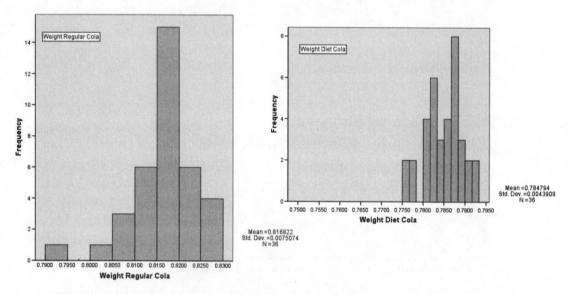

Chapter 2

1.   The histogram for weights of Regular Cola and weights of Diet Cola.

The histogram for weights of Regular Cola much less spread out around the center of about *0.8170*, while the weights of the Diet Cola are centered near *0.7848*. The histograms have different centers and different shapes. The histogram for weights of regular cola shows a possible outlier as well.

2. The histogram for measured weights of the bears.

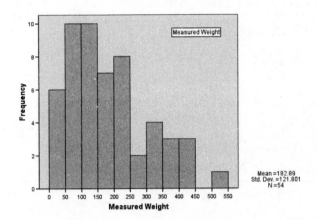

3. The side-by-side boxplots for the weights of Regular Cola, Diet Cola, Regular Pepsi, and Diet Pepsi are given below. Diet colas weigh less than regular colas.

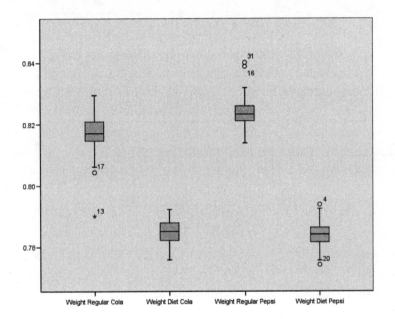

4. There is a high energy consumption when the weather is cold and when the weather is hot.

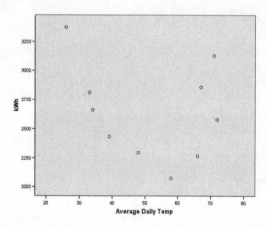

Chapter 3

1. a. The mean and standard deviation of the weights of Regular Cola, Diet Cola, Regular Pepsi, and Diet Pepsi are given below. Regular Cola and Regular Pepsi weigh more and have larger standard deviations than Diet Cola and Diet Pepsi.

		Weight Regular Cola	Weight Diet Cola	Weight Regular Pepsi	Weight Diet Pepsi
N	Valid	36	36	36	36
	Missing	0	0	0	0
Mean		.816822	.784794	.824103	.783858
Std. Deviation		.0075074	.0043909	.0057011	.0043616

b. The 5-number summary for the weights of Regular Cola, Diet Cola, Regular Pepsi, and Diet Pepsi are given below.

Statistics

		Weight Regular Cola	Weight Diet Cola	Weight Regular Pepsi	Weight Diet Pepsi
N	Valid	36	36	36	36
	Missing	0	0	0	0
Median		.817100	.785200	.823300	.784150
Minimum		.7901	.7758	.8139	.7742
Maximum		.8295	.7923	.8401	.7938
Percentiles	25	.814475	.782200	.821100	.781550
	50	.817100	.785200	.823300	.784150
	75	.821000	.787900	.826000	.786475

2. The required statistics (rounded to two decimal places) for age are shown below
   a) Mean = *54.60*   b) Median = *54.00*   c) Mode = *51.00* (not unique)
   d) Range = *27.00*   e) Stand. dev. = *6.13*   f) Variance = *37.58*
   g) Q_1 = *51.00* Q_3 = *57.00*   h) P_{10} = *46.4*

3. The Descriptive Table is shown below.

Statistics

interval

N	Valid	12
	Missing	0
Mean		91.25
Std. Error of Mean		2.556
Median		93.50
Mode		92[a]
Std. Deviation		8.854
Variance		78.386
Range		33
Minimum		65
Maximum		98
Sum		1095
Percentiles	25	90.50
	50	93.50
	75	95.75

a. Multiple modes exist. The smallest value is shown

4. By comparing the means and medians, the eastbound stowaways appear to be somewhat younger.

Statistics

		Eastbound	Westbound
N	Valid	10	10
	Missing	0	0
Mean		21.30	29.50
Median		20.00	25.00
Mode		20[a]	21[a]

a. Multiple modes exist. The smallest value is shown

Chapter 4

1. The histogram of *100* Uniform random numbers on the interval (*0, 10*) is shown below. Type RV.*UNIFORM(0, 10)* into the **Numeric Expression** box.

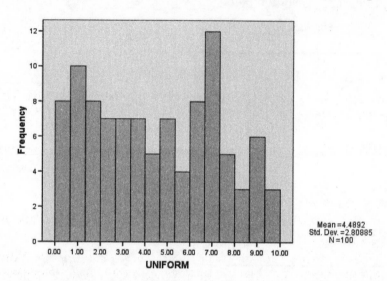

2. The histogram of *100* Uniform random digits between *1* and *10*, inclusive is shown below. Type *TRUNC(RV.UNIFORM(1, 11))* into the **Numeric Expression** box. The histogram was made by choosing **Graphs > Bar…**

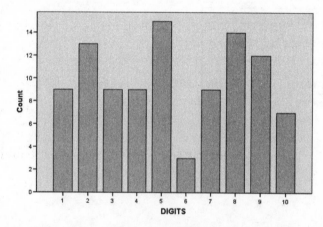

3. Answers will vary since the result depends on simulation. Type *RV.BINOM(20, 0.6)* into the **Numeric Expression** box. The first *11* appears in the sorted variable **binomial** at *253*. Therefore, *747* simulations of the *1000*, or *74.7%* of the simulations, had more girls than boys.

4. Answers will vary since the result depends on simulation. Make three variables, *die1*, *die2*, and *die3*, by typing *TRUNC(RV.UNIFORM(1, 7))* into the **Numeric Expression** box. Make the variable *sum* by typing *die1 + die2 + die3* into the **Numeric Expression** box. Sort the variable *sum*; the number of *10*'s is *15*; therefore the simulated probability is *15 / 100* or *0.15*.

Chapter 5

1. Type *1 - CDF.BINOM(11, 20, 0.25)* in the **Numeric Expression** box. The probability is *0.0009354*.

2. Type *CDF.BINOM(1, 24, 0.04)* in the **Numeric Expression** box. The probability is *0.7508*.

3. The probability of tossing a *1* is *1/6*. Store *1/6* in variable *p*.
    a. Type *CDF.BINOM(20, 100, p)* in the **Numeric Expression** box and *prob1* in the Target Variable box. The probability is *0.8481*.
    b. Type *CDF.BINOM(20, 100, p) – CDF.BINOM(19, 100, p)* in the **Numeric Expression** box and type *prob2* in the Target Variable box. The probability is *0.06786*.

4. The mean value is *11/365 = 0.03*.
    a. Type *CDF.POISSON(0, 0.03)* in the **Numeric Expression** box. The probability is *0.9704*.
    b. Type *1 - CDF.POISSON(0, 0.03)* in the **Numeric Expression** box. The probability is *1 - 0.9704 = 0.0296*.

5. The mean value is *10,000 $\times 0.001 = 10$*. Type *1 - CDF.POISSON(20, 10)* in the **Numeric Expression** box. The probability is *0.00159*.

6. The mean value is *93 / 100 = 0.93*. Type *CDF.POISSON(1, 0.93) - CDF.POISSON(0, 0.93)* in the **Numeric Expression** box. The probability is *0.3669*.

Chapter 6

1. Assuming that women's weights are N(*143, 29*).
   a. Type *CDF.NORMAL(120, 143, 29)* in the **Numeric Expression** box to calculate the probability, P($x < 120$) = *0.2139*.
   b. Type *1 - CDF.NORMAL(135, 143, 29)* in the **Numeric Expression** box to calculate the probability, P($x > 135$) = *0.6087*.
   c. Type *CDF.NORMAL(152, 143, 29) - CDF.NORMAL(128, 143, 29)* in the **Numeric Expression** box to calculate the probability, P(*128 < x < 135*) = *0.3194*.

2. Assuming that women's weights are N(*143, 29*).
   a. Type *IDF.NORMAL(0.90, 143, 29)* in the **Numeric Expression** box to calculate the 90th percentile of the distribution, P_{90} = *180.16*.
   b. Type *IDF.NORMAL(0.50, 143, 29)* in the **Numeric Expression** box to calculate the median of the distribution, P_{50} = *143*. The median and the mean of Normal distributions are equal.

3. You will first need to use the syntax editor (see Figure 6 – 1) to generate 100 cases, each having an arbitrary value of 0. Answers will vary because of randomness. Type *RV.NORMAL(123, 4)* into the **Numeric Expression** box to determine the random sample. The mean of both samples were close to population mean *123* (*122.37* and *122.86*, respectively). The standard deviations of both samples were near to the population standard deviation *4* (*4.07* and *4.15*, respectively). This should be true from sample to sample.

4. The exact binomial distribution is Bin(*580, 0.25*) and the approximate normal distribution is N(*145, 10.428*). Using the Normal approximation the probability is found by typing *1 - CDF.NORMAL(152, 145, 10.428)* in the **Numeric Expression** box. The exact probability is found by typing *1 - CDF.BINOM(151, 580, 0.25)* in the **Numeric Expression** box. The approximate probability is P($x > 152$) = *0.2510*. The exact probability is P($x > 152$) = *0.2650*. The probabilities are very close; the approximate value (not using the continuity correction, which would improve the approximation) is about *5.3%* in error.

5. Make a histogram and a Normal quantile plot to determine Normality.
    a. The histogram for weights of Regular Cola shows a possible outlier. The Normal quantile plot is substantially away from a straight line because of this possible outlier. Regular Cola does not have a normal distribution.

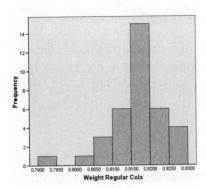 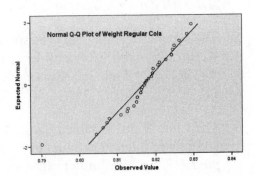

    b. The histogram for weights of Diet Cola shows no outliers and is fairly symmetric. The Normal quantile plot looks very straight and so we conclude that Diet Cola does have a normal distribution.

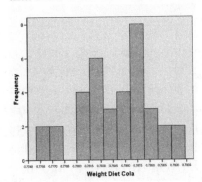 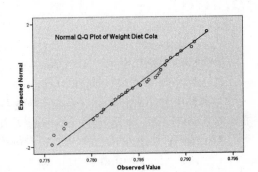

    c. and
    d. The histograms for weights of Regular Pepsi and Diet Pepsi show no outliers and are fairly symmetric. The Normal quantile plots looks very straight and so we conclude that Regular Pepsi and Diet Pepsi have a normal distribution.

Chapter 7

1. A *98%* confidence interval for the mean amount of Tar is (*9.98, 14.23*).

2. A *95%* confidence interval for the variance in the amount of Tar is (*13.51, 39.24*). The lower limit is found by typing *(29 - 1) * 21.453 / IDF.CHISQ(0.975, 28)* into the **Numeric Expression** box.

3. A *95%* confidence interval for the mean amount of Carbon Monoxide is (*10.68, 14.08*).

4. A *90%* confidence interval for the variance in the amount of Carbon Monoxide is (*13.52, 33.01*). The lower limit is found by typing *(29 - 1) * 19.958 / IDF.CHISQ(0.95, 28)* into the **Numeric Expression** box.

5. The sample size necessary to estimate the mean amount of Carbon Monoxide within 1mm with *95%* confidence is found by typing *(IDF.NORMAL(0.975, 0, 1) * 4.467 / 1)**2* into the Numeric Expression box. The necessary sample size is *77*.

6. The value of the variable *phat* is *28/47 = 0.5957*. The margin of error for this confidence interval is found by typing *IDF.NORMAL(0.98, 0, 1) * SQRT(phat*(1-phat)/47)* into the **Numeric Expression** box. The **margin of error** is *0.15*. The confidence interval is *phat* ± **margin of error**, or (*0.45, 0.74*).

Chapter 8

1. Test the claim that

    $$H_0: p = 0.10 \qquad H_1: p > 0.10$$

    at the *0.05* significance level. The value of variable *phat* is *14/59 = 0.2373*. The test statistic z is found by typing *(phat - 0.10) / SQRT(0.10*0.90/59)* into the **Numeric Expression** box. The test statistic **z** is *3.52*. The P-value for this test $P(z > 3.52)$ is found by typing *1 - CDF.NORMAL(z, 0, 1)* into the **Numeric Expression** box. The P-value is *0.0002*. Therefore, reject the null hypothesis and conclude that the proportion of blue M&Ms is more than *10%*.

2. Test the claim that

    $$H_0: \mu = 0.9085 \qquad H_1: \mu > 0.9085$$

    at the *0.10* significance level. The P-value is *0.795/2 = 0.3975*. Therefore, we fail to reject the null hypothesis and conclude that the gummy bears do not weigh more than *0.9085* grams.

3. Test the claim that

    $$H_0: \mu = 3.14 \qquad H_1: \mu \neq 3.14$$

    at the *0.05* significance level. The P-value is *0.217*. Therefore, we fail to reject the null hypothesis and conclude that the mean of the variable *ratio* is π.

4. Test the claim that

$$H_0: \mu = 40 \qquad H_1: \mu > 40$$

at the *0.01* significance level. The P-value is *0.023*. Therefore, we fail to reject the null hypothesis and conclude that the mean of the nicotine content is not greater than *40*.

5. Test the claim that

$$H_0: \sigma = .051 \qquad H_1: \sigma < .051$$

at the *0.05* significance level. The X^2 value is 13.6436. The P-value is *0.634*. Therefore, we fail to reject the null hypothesis and conclude that the standard deviation of the weights of the cola cans is not less *0.051*.

Chapter 9

1. a. Test the claim that

$$H_0: p_{2000} - p_{2001} = 0 \quad H_1: p_{2000} - p_{2001} > 0$$

at the *0.01* significance level. The value of variable **phat1** is *83/247 = 0.3360*, **phat2** is *89 / 258 = 0.3450*, and **pbar** is *(83+ 89) / (247+ 258) = 0.3406*. The test statistic *z* is found by typing *(phat1 - phat2) / SQRT(pbar * (1-pbar) / 247 + pbar * (1-pbar) / 253)* into the **Numeric Expression** box. The test statistic **z** is *-0.21*. The P-value for this test P(*z* > *-0.21*) is found by typing *1 - CDF.NORMAL(z, 0, 1)* into the **Numeric Expression** box. The P-value is *0.5834*. Therefore, fail to reject the null hypothesis and conclude that the proportion of video replays reversed has not changed.

   b. The margin of error for this confidence interval is found by typing *IDF.NORMAL(0.975, 0, 1) * SQRT(phat1 * (1-phat1) / 247 + phat2 * (1-phat2) / 253)* into the **Numeric Expression** box. The **margin of error** is *0.08*. The confidence interval is **phat1 – phat2** $\pm$ **margin of error**, or (*-0.09*, *0.07*).

2. Test the claim that

$$H_0: \mu_{Orange} - \mu_{Brown} = 0 \quad H_1: \mu_{Orange} - \mu_{Brown} \neq 0$$

at the *0.01* significance level. The P-value of the test associated with equal variance is *0.673*. Therefore, we fail to reject the null hypothesis and conclude that the mean weights of orange and brown M&Ms are the same.

3. A *90%* confidence interval for the difference in the mean weight of orange and brown M&Ms is *(-0.030, 0.050)*.

4. Test the claim that

$$H_0: \mu_{Blue} - \mu_{Not\ Blue} = 0 \quad H_1: \mu_{Blue} - \mu_{Not\ Blue} \neq 0$$

   at the *0.05* significance level. The P-value of the test associated with equal variance is *0.958*. Therefore, we fail to reject the null hypothesis and conclude that the mean weights of blue and not blue M&Ms are the same.

5. A *98%* confidence interval for the difference in the mean weight of blue and not blue M&Ms is *(-0.028, 0.027)*.

6. Test the claim that

$$H_0: \sigma^2_{Orange} = \sigma^2_{Brown} \quad H_1: \sigma^2_{Orange} \neq \sigma^2_{Brown}$$

   at the *0.05* significance level. Use the Explore procedure to determine:

	Orange	Brown
n	25	8
$\bar{x}$	.85780	.84775
s	.050104	.079500

   Compute $F = .3972$. Then $P(P > .3972) = 1 - CDF.F(F, 24, 7) = .9569$ Since the P-value is greater than the significance level, we fail to reject the null hypothesis and conclude that the variation in the samples is the same.

7. Let d = $6^{th} - 13^{th}$. Test the claim

$$H_0: \mu_d = 0 \quad H_1: \mu_d \neq 0$$

   The P-value for this two-tailed test is .042 which is less than the significance level; so, there is sufficient evidence to reject the null hypothesis and conclude that the mean difference in the admissions is not the same.

8. A 90% confidence interval for the mean difference in admissions between non-Friday 13 and Friday 13 is *(-5.801, -.856)*.

9. Test the claim that

$$H_0: \sigma^2_{Student} = \sigma^2_{Faculty} \quad H_1: \sigma^2_{Student} > \sigma^2_{Faculty}$$

   at the *0.05* significance level. The test statistic F is found by typing *3.67**2 / 3.65**2* into the **Numeric Expression** box. The test statistic F is *1.01*. The P-value for this test P(F > *1.01*) is found by typing *1 - CDF.F(F, 217, 152)* into the **Numeric Expression** box. The P-value is *0.4744*. Therefore, fail to reject the null hypothesis and conclude that there is insufficient information to claim that the variability of the age of student cars is greater than the variability of faculty cars at the college.

Chapter 10

1. Pearson's correlation coefficients between **weight** and **age**, **headlen**, **headwth**, **neck**, **length**, and **chest** are given below.

Correlations		Age	Length of Head	Width of Head	Distance around neck	Length of body	Distance around the chest
Measured Weight	Pearson Correlation	.749**	.834**	.783**	.934**	.864**	.963*
	Sig. (2-tailed)	.000	.000	.000	.000	.000	.000
	N	54	54	54	54	54	54

**. Correlation is significant at the 0.01 level (2-tailed).

2. The variable **month** has Nominal scale and therefore is inappropriate for Pearson's correlation coefficient.

3. The regression equation is **weight = –264.48 + 12.54 chest**. The coefficient of determination (percent of variation of **weight** explained by **chest**) is **92.8%**. (See the SPSS output below.)

Model Summary

Model	R	R Square	Adjusted R Square	Std. Error of the Estimate
1	.963a	.928	.926	33.078

a. Predictors: (Constant), Distance around the chest

Coefficients^a

Model		Unstandardized Coefficients		Standardized Coefficients	t	Sig.
		B	Std. Error	Beta		
1	(Constant)	-264.481	17.902		-14.774	.000
	Distance around the chest	12.544	.486	.963	25.819	.000

a. Dependent Variable: Measured Weight

4. A **98%** prediction interval for the weight of a bear with a chest measurement of **60** inches is **(403.2, 573.2)**.

5. The multiple regression equation is **weight = -267.06 + 9.29 chest + 5.77 neck**. The adjusted coefficient of determination is **0.934**. The percent of variation of weight explained by chest and neck is **93.7%**.

Model Summary^b

Model	R	R Square	Adjusted R Square	Std. Error of the Estimate
1	.968a	.937	.934	31.252

a. Predictors: (Constant), Distance around neck, Distance around the chest

b. Dependent Variable: Measured Weight

Coefficients[a]

Model		Unstandardized Coefficients		Standardized Coefficients		
		B	Std. Error	Beta	t	Sig.
1	(Constant)	-267.060	16.941		-15.764	.000
	Distance around the chest	9.291	1.293	.713	7.188	.000
	Distance around neck	5.771	2.143	.267	2.693	.010

a. Dependent Variable: Measured Weight

6. A 95% prediction interval for the weight of a bear with a chest measurement of *60* inches and neck measurement 30 inches is (*393.8, 533.1*).

Chapter 11

1. Test the claim that

H_0: $p_1 = p_2 = p_3 = \dots p_{10} = 0.10$ H_1: at least one of the *10* p_k are not equal to *0.10*

at the *0.05* significance level. The P-value of the test is *0.898*. Therefore, we fail to reject the null hypothesis and conclude that the digits are all equally likely.

2. a. A binomial probability distribution with *n = 3* and *p = 1/3* has probabilities

x	*0*	*1*	*2*	*3*
P(x)	*0.2963*	*0.4444*	*0.2222*	*0.0370*

   b. Test the claim that

   H_0: $p_0 = 0.2963$, $p_1 = 0.4444$, $p_2 = 0.2222$, $p_3 = 0.0370$
   H_1: at least one of the *10* p_k are not equal to *0.10*

   at the *0.05* significance level. The P-value of the test is *0.000*. Therefore, we reject the null hypothesis and conclude the sample does not come from a binomial distribution with *n = 3* and *p = 1/3*.

3. Test the claim that

   H_0: Confidence in police and Gender are independent
   H_1: Confidence in police and Gender are not independent

   at the *0.05* significance level. The P-value of the Pearson Chi-Square test is *0.334*. Therefore, we fail to reject the null hypothesis and conclude the Confidence in police and Gender are independent.

4. Test the claim that

   H_0: Political philosophy and News affiliate are independent
   H_1: Political philosophy and News affiliate are not independent

   at the *0.05* significance level. The P-value of the Pearson Chi-Square test is *0.998*. Therefore, we fail to reject the null hypothesis and conclude the Political philosophy of an individual is independent of the affiliated news station the individual watches.

5. Test the claim that

H_0: $p_{B,G} = 0.5625$, $p_{B,R} = 0.1875$, $p_{R,G} = 0.1875$, $p_{R,R} = 0.0625$
H_1: at least one of the 4 probability statements is untrue

at the *0.05* significance level. The P-value of the test is *0.484*. Therefore, we fail to reject the null hypothesis and conclude the results are consistent with the expected proportion.

Chapter 12

1. Test the claim that

H_0: $\mu_{subcompact} = \mu_{compact} = \mu_{midsize} = \mu_{full\text{-}size}$
H_1: The means are not all equal

at the *0.05* significance level. The P-value of the F test is *0.296*. Therefore, we fail to reject the null hypothesis and conclude that the four sizes of cars have the same mean numbers of Chest Deceleration injuries.

2. Test the claim that

H_0: $\mu_{subcompact} = \mu_{compact} = \mu_{midsize} = \mu_{full\text{-}size}$
H_1: The means are not all equal

at the *0.05* significance level. The P-value of the F test is *0.413*. Therefore, we fail to reject the null hypothesis and conclude that the four sizes of cars have the same mean numbers of Left Femur Load injuries.

3. Test the claim that

H_0: $\mu_{red} = \mu_{orange} = \mu_{yellow} = \mu_{brown} = \mu_{blue} = \mu_{green}$
H_1: The means are not all equal

at the *0.05* significance level. The P-value of the F test is *0.769*. Therefore, we fail to reject the null hypothesis and conclude that the different colors of M&Ms all have the same mean weight. The company does not have a problem.

4. At the *0.05* significance level, there is no effect due to gender-age interaction (P-value = *0.702*); there is no effect due to gender (P-value = *0.762*); and there is no effect due to age (P-value = *0.702*).

Chapter 13

1. Test the claim that

    H_0: the medians of the weight of discarded metal and discarded paper are equal

    H_1: The medians are not all equal

    at the 0.05 significance level. All 62 households had more metal than paper discarded by weight; the P-value of the Sign test is 0.000. Therefore, we reject the null hypothesis and conclude that households discard different median weights of metal and paper.

2. Test the claim that

    H_0: the medians of the weight of discarded metal and discarded paper are equal

    H_1: The medians are not all equal

    at the 0.05 significance level. All 62 households had more metal than paper discarded by weight; the P-value of Wilcoxon Signed-Ranks Z test is 0.000. Therefore, we reject the null hypothesis and conclude that households discard different median weights of metal and paper.

3. Test the claim that

    H_0: The weights of orange and brown M&Ms have the same distribution

    H_1: The two samples come from different distributions

    at the 0.05 significance level. The P-value of the Wilcoxon Rank-Sum test is 0.726. Therefore, we fail to reject the null hypothesis and conclude that orange and brown M&Ms have the same distribution.

4. Test the claim that

    H_0: $\mu_{subcompact} = \mu_{compact} = \mu_{midsize} = \mu_{full\text{-}size}$

    H_1: The means are not all equal

    at the 0.05 significance level. The P-value of the Kruskal-Wallis Chi-Square test is 0.237. Therefore, we fail to reject the null hypothesis and conclude that the four sizes of cars have the same mean numbers of Chest Deceleration injuries.

5. Spearman's rank correlation coefficients between *weight* and *age*, *headlen*, *headwth*, *neck*, *length*, and *chest* are given below.

Correlations

Spearman's rho		Age	Length of Head	Width of Head	Distance around neck	Length of body	Distance around the chest
Measured Weight	Correlation Coefficient	.854**	.915**	.853**	.959**	.941**	.984**
	Sig. (2-tailed)	.000	.000	.000	.000	.000	.000
	N	54	54	54	54	54	54

**. Correlation is significant at the .01 level (2-tailed).

6. The P-value for the Runs Test for Randomness is *0.000*; this indicates that the data are not random. Looking at the data, we see that 16 values are greater than or equal to the median and 14 are less than the media and that all the values greater than the median are in later years. This indicates that the Dow Jones Industrial Average is tends to increase over time.

Chapter 14

1. The process appears to be within statistical control. There is no obvious trend or shift in the chart and there are only two data values outside the control limits. See the graph below.

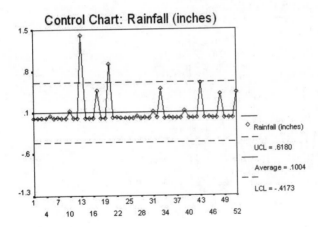

2. The process appears to be out of statistical control. Although, there is no obvious trend or shift in the chart and there are only two data values outside the control limits; there is a run of eight negative values. See the graph below.

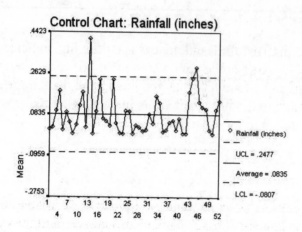

3. The process appears to be out of statistical control. Although, there is no obvious trend or shift in the chart and there are no long runs; there are *17* data values outside the control limits. See the graph below.

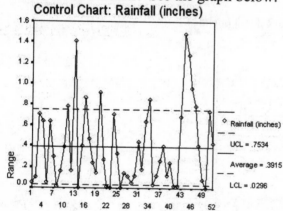

4. The process appears to be within statistical control. There is no obvious trend or shift in the chart and there are only four data values outside the control limits. See the graph below.

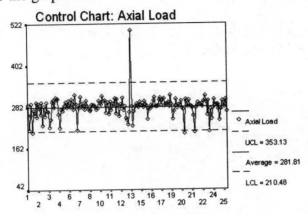

5. The means chart displays no obvious trend or shifts in the chart and there are no data values outside the control limits. There is also no run of eight values to one side of the mean. The mean of the process is within statistical control. The range chart shows no obvious trend or shifts in the chart and there is only one data value outside the control limits. There is also no run of eight values to one side of the mean. The spread of the process is within statistical control. See the graphs below.

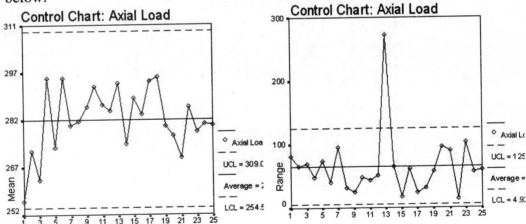

6. The process appears to out of statistical control. There is a downward trend, and there are nine consecutive points lying below the centerline. See the graph below.

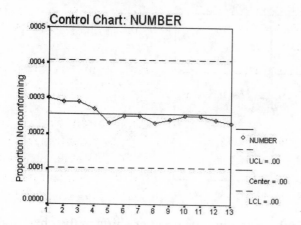

7. The process is out of control. All the values are outside the control limits for the process. See the graph below.

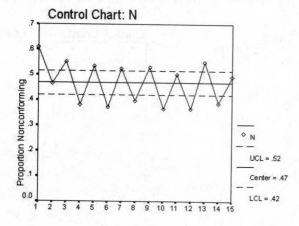

INDEX